Elements of Wavelets for Engineers and Scientists

Elements of Wavelets for Engineers and Scientists

DWIGHT F. MIX
Department of Electrical Engineering
University of Arkansas

KRAIG J. OLEJNICZAK
Department of Electrical and Computer Engineering
Valparaiso University

A JOHN WILEY & SONS, INC., PUBLICATION

Published by John Wiley & Sons, Inc., Hoboken, New Jersey.
Published simultaneously in Canada.

Library of Congress Cataloging-in-Publication Data Is Available

ISBN 0-471-46617-4

10 9 8 7 6 5 4 3 2 1

Contents

Preface vii

1. Functions and Transforms 1
1.1. Wavelet Transform 1
1.2. Transforms 5
1.3. Power and Energy Signals 9
1.4. Deterministic and Random Signals 16
1.5. Fourier and Haar Transforms 18

2. Vectors 25
2.1. Vector Space 26
2.2. Metric Space 31
2.3. Norm 36
2.4. Inner Product 39
2.5. Orthogonality 45

3. Basis and Dimension 47
3.1. Linear Independence 48
3.2. Basis 51
3.3. Dimension and Span 54
3.4. Reciprocal Bases 56

4. Linear Transformations 65
4.1. Component Vectors 65
4.2. Matrices 69

5. Sampling Theorem 80
5.1. Nyquist Rate 80
5.2. Nonperiodic Sampling 88
5.3. Quantization and Pulse Code Modulation 90
5.4. Companding 93

6. Multirate Processing 95
6.1. Downsampling 95
6.2. Upsampling 103
6.3. Fractional Rate Change 104
6.4. Downsampling and Correlation 110
6.5. Upsampling and Convolution 116

7. Fast Fourier Transform 121
7.1. Discrete-Time Fourier Series 121
7.2. Matrix Decomposition View 125
7.3. Signal Flow Graph Representation 128
7.4. Downsampling View 139

8. Wavelet Transform 145
8.1. Scaling Functions and Wavelets 145
8.2. Discrete Wavelet Transform 162

9. Quadrature Mirror Filters 170
9.1. Allpass Networks 170
9.2. Quadrature Mirror Filters 181
9.3. Filter Banks 186

10. Practical Wavelets and Filters 195
10.1. Practical Wavelets 195
10.2. The Magic Part 201
10.3. Other Wavelets 208
10.4. Matrix of Transformation 213

11. Using Wavelets 219
11.1. Top-Down Approach 219
11.2. Pattern Recognition 227
11.3. Hidden Singularities 230
11.4. Data Compression 232

Index 235

Preface

Why another book on wavelets? There are already over 150 such books on the market, with more being written every day. Workers in the field judge some of these "excellent".

Over a year ago, we decided to include a chapter or two on wavelets in a textbook we were writing. Our plan was to study a few articles and texts to better understand wavelets before writing on the subject. Little did we know that the literature was so abstruse. Despite diligent searching, we have yet to find a book on the subject that contains a single worked out numerical example. Rare is the book on wavelets that contains an example of any kind. Many hours of study convinced us that there is a need for an introductory book on wavelets for scientists and engineers that act as the prerequisite to the other 150 mentioned above.

This book is for the rest of us, the non-mathematicians who want to understand wavelets. Our primary goal is to maximize your probability of success in the comprehension and retention of a very mathematically sophisticated topic. Our secondary goal is to conceptually prepare you for the more mathematically rigorous wavelet texts that lie ahead. The presentation is as simple as possible, but to paraphrase Einstein, "it should not be simpler." The present state of the art is such that some of the more esoteric mathematical concepts cannot be avoided. This book attempts to provide an explanation of this esoteric math with plenty of examples, diagrams, and drill problems. There are 88 examples, 145 diagrams, and 40 drill problems in this short book. All of the examples are worked out, and most are numerical.

The reader should bring a math background to the table that is typical of most holders of the Bachelor of Science degree, i.e., some background in calculus, power series, and set theory with a smattering of "higher mathematics" related to algebra or geometry. Prior experience with transforms will also help. As with any new topic, the effort required to understand wavelets is inversely proportional to familiarity with this background material.

We wish you well on your "wavelet" excursion.

Fayetteville, Arkansas — Dwight F. Mix, Ph.D., P.E.
Valparaiso, Indiana — Kraig J. Olejniczak, Ph.D., P.E.

March 9, 2003

Chapter1
Functions and Transforms

The goal of this text is to provide the reader with an explanation of the fast Fourier and wavelet transforms. The goal in this chapter is to provide the reader with an explanation of the general idea behind transforms and to introduce some necessary terminology. The topics in this chapter include transforms, fields, and signal classification. The fast Fourier and wavelet transforms are covered in more detail later in the book.

Chapter Goals: After completing this chapter, you should be able to do the following:

- Define *transform*.
- Determine if a given signal is a power or energy signal.
- For a given discrete-time signal of length 4, calculate both the discrete-time Fourier series and the Haar transforms.

1.1. Wavelet Transform

There are two ways to determine the wavelet transform. Most transforms have only one way, a formula. For example, the Laplace transform of a time function $v(t)$ is calculated by the formula

$$V(s) = \int_{-\infty}^{\infty} v(t)e^{-st}\, dt$$

There is a similar formula for the wavelet transform, but we want to show you the other view in this introduction. This second view is the way the wavelet transform is calculated in practice.

Start with samples of a continuous-time signal over some interval. Let us say that the interval is long enough to include 1024 samples. Supply this discrete-time signal to two digital filters in parallel. Figure 1.1 shows the process, where H_0 is a low-pass filter and H_1 is a high-pass filter. The input is $s(n)$, and $c(n)$ and $d(n)$ are the output terms. The $(\downarrow 2)$ symbol represents downsampling by 2, which means that we throw away every other sample. We keep the 1st, 3rd, 5th, ... samples and throw away the 2nd, 4th, 6th, ... samples. In this scheme $c(n)$ contains 512 samples, as does $d(n)$.

The 512 $d(n)$ terms represent part of the transform (in fact, half of the transform). The other half is extracted from $c(n)$ by repeating the filter-downsample operation. Figure 1.2 shows exactly what this means. Note that the first filter-downsample terms are c_9 and d_9. This is because they contain 512 samples and $2^9 = 512$. Similarly, the second-stage output is labeled c_8 and d_8 because it contains 256 terms. The 256 d_8 terms represent an additional one-fourth of the transform. Now you can guess how to calculate the additional terms. Repeating the filter-downsample operation once again on the c_8 terms generates c_7 and d_7, each containing 128 terms.

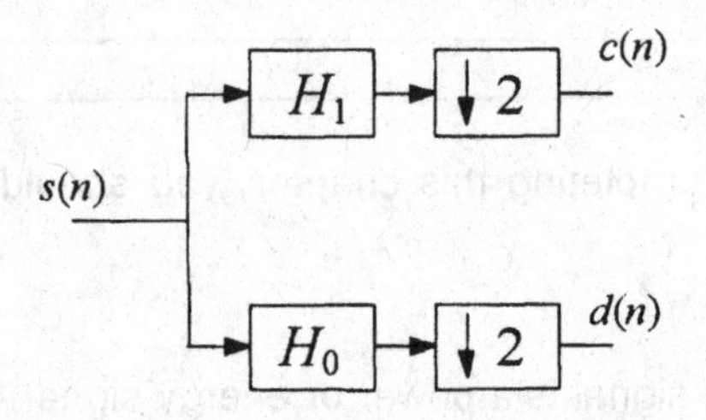

Figure 1.1. The first stage of the transform.

The wavelet transform contains the same number of terms as the signal, 1024 in this example. The terms in the transform consist of d_9 (512 samples), d_8 (256 samples), d_7 (128 samples), d_6 (64 samples), d_5 (32 samples), d_4 (16 samples), d_3 (8 samples), d_2 (4 samples), d_1 (2 samples), d_0 (1 sample), plus the one term in c_0.

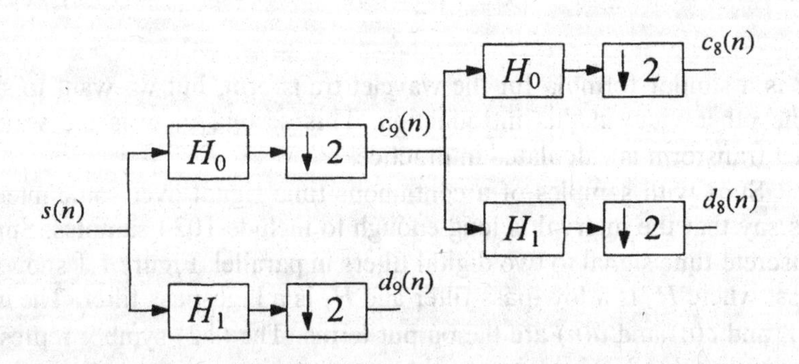

Figure 1.2. The first two stages of the transform.

To find the inverse transform, reverse this process. Start with the two terms c_0 and d_0. Supply these to two digital filters in parallel, as shown in Fig. 1.3. The (↑2) symbol represents upsampling by 2, which means that we insert zeros between samples. Starting with a signal of length 4 given by {1, 2, −1, 3}, the resulting upsampled signal is {1, 0, 2, 0, −1, 0, 3, 0}, which has length 8. Initially, there is one sample at the input of each upsample-filter branch, the output will contain two samples. For proper filters H_2 and H_3 the output of the first summer will be $c_1(n)$.

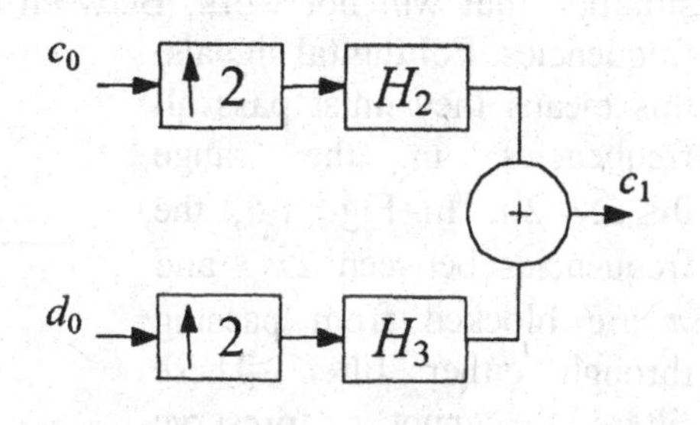

Figure 1.3. The Inverse Transform.

Figure 1.4 shows the first two stages of the inverse transform. This process continues until the last output is $s(n)$, the original signal. Note that we must supply $c_0(n)$, $d_0(n)$, $d_1(n)$, $d_2(n)$,..., $d_9(n)$ at successive stages of this process. These are the terms that we saved and called the transform in the forward process.

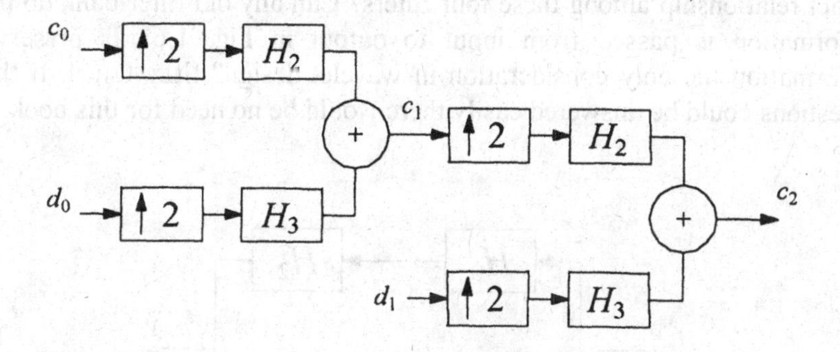

Figure 1.4. The first two stages of the inverse transform.

Little has been said about the filters in this process, except that H_0 is low-pass and H_1 is high-pass. What characteristics must these filters have, and what is their relationship one to another? We can answer some of these questions now by knowing that $c_0(n)$, $d_0(n)$, $d_1(n)$, $d_2(n)$, ..., $d_9(n)$ must contain the same information as $s(n)$. The remainder of this book should provide enough information to answer most of these questions.

Consider the relationship between H_0 and H_1. Figure 1.5 shows the situation that will not work. Between them, the two filters must pass all frequencies. For digital signals, this means they must pass all frequencies in the range $0 \le \Omega < 2\pi$. In Fig. 1.5, the frequencies between $2\pi/3$ and π are blocked from passing through either filter. These filters cannot preserve information.

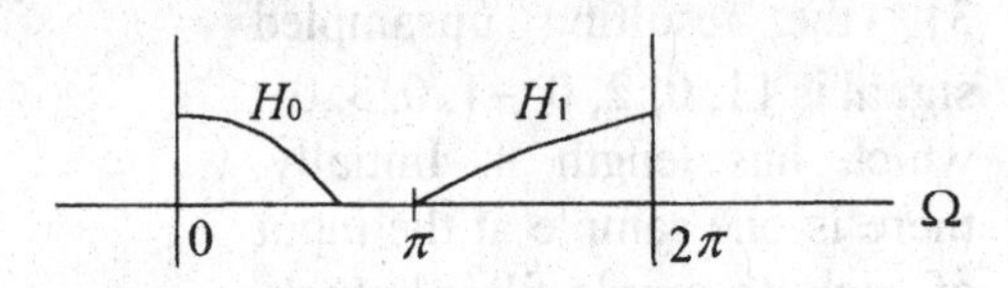

Figure 1.5. Two frequency responses.

Figure 1.6 shows a filter bank that depicts the relationship between the four filters without the (↓2) and (↑2) operations. Consider the relationship between H_0 and H_2, or the top branch in Fig. 1.6. If the output y contains all the information in the input v, then H_2 must also be a low-pass filter. Suppose, to the contrary, that H_0 is low-pass and H_2 is high-pass. Then all frequency components will be blocked from passing through the top branch. Similarly, if H_1 is high-pass, then H_3 must also be high-pass.

This still leaves many questions to be answered. What must be the exact relationship among these four filters? Can any old filter bank do if all information is passed from input to output in Fig. 1.6? Is preserving information the only consideration in wavelet design? Stay tuned. If these questions could be answered easily there would be no need for this book.

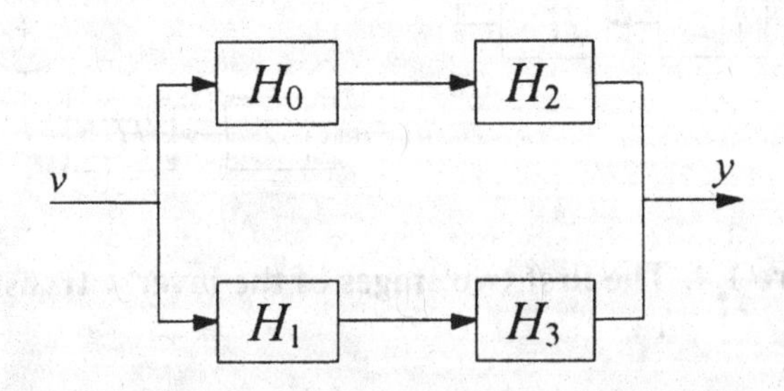

Figure 1.6. Filter bank with all four filters.

1.2. Transforms

A transform is a special type of function. To understand transforms, one must first understand the concept of a function. The reader will recall that a *function* is a relationship between two sets (called the *domain* and *codomain*), and this relationship must satisfy two conditions. First, every element in the domain must correspond to some element in the codomain. Second, no two elements in the codomain can correspond to the same element in the domain. These two requirements can be combined by saying that for each element in the domain there corresponds exactly one element in the codomain.

Figure 1.7 shows a map or function. We write this $f: A \to B$ and call the function f, the domain A, and the codomain B. Notice that each element in the domain A is matched up with something in B (the first requirement) and that no two elements in B match up with the same element in A (the second requirement.) It is OK for two different elements in A to match with the same element in B.

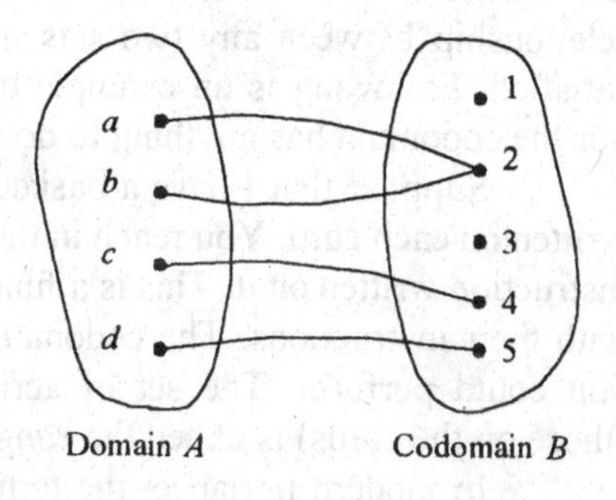

Figure 1.7. The function $f: A \to B$.

The domain is the set $\{a, b, c, d\}$ and the codomain is the set $\{1, 2, 3, 4, 5\}$. The *range* is the set $\{2, 4, 5\}$. The range consists of the elements in B that have an ancestor in A, and they form a subset of B. The range and codomain can be different sets, with the range a subset of the codomain. This distinction between the codomain and range occurred recently in mathematics. Previously, both sets were called the range, which led to some confusion.

Another definition of function is a set of ordered pairs where the first element is not repeated. Consider the following three sets of ordered pairs.

$(a,2)$	$(a,2)$	$(a,1)$
$(b,2)$	$(b,2)$	$(b,2)$
$(c,4)$	$(a,4)$	$(c,4)$
$(d,5)$	$(d,5)$	$(d,5)$

The first column represents the function f in Fig. 1.7. The second column is not a function because a is repeated. The third column is a function.

Definition 1.1. A *function* $f: A \to B$ is a relationship between two sets A and B such that for each element in the first set there corresponds exactly one element in the second set.

Other names for *function* are *map* and *mapping*. In the beginning, the concept of function applied to numbers. The domain and codomain consisted of numbers. These are the functions for which graphs are drawn and formulas are written. Now the concept of function applies to the relationship between any two sets if the two properties for a function are satisfied. Following is an example of a function where neither the domain nor the codomain has anything to do with numbers.

Suppose that I have a basket full of cards, and that an instruction is written on each card. You reach in the basket, select a card, and carry out the instruction written on it. This is a function. The domain is the basket of cards with their instructions. The codomain is the set of all possible actions that you could perform. The set of actions that might actually be performed (those on the cards) is called the *range*, and this is a subset of the codomain.

In modern parlance, the term *range* refers to those elements in the codomain that have an ancestor in the domain. Figure 1.8 shows the domain and codomain as baskets. All values of y fall into the subset of the codomain called the range after being ejected from the box labeled f. The function $y = x^2$ provides a specific example. If the domain and codomain are the set of all real numbers, the range is the set of all non-negative real numbers. Therefore, in this example the range is a proper subset of the codomain. (*Proper* means that some elements in the codomain are not in the range.)

Now for the idea of a transform. The functions of interest to us here also have no numbers in either the domain or codomain. These special functions go by either of two names, *transform* or *operator*. They are functions whose domain and range are themselves sets of functions. If Fig. 1.8 represents a transform, all baskets contain functions instead of numbers. For example, in the Fourier transform the domain contains time functions and the codomain contains frequency functions. The formula for the Fourier transform is given by

$$V(\omega) = \int_{-\infty}^{\infty} v(t) e^{-j\omega t}\, dt$$

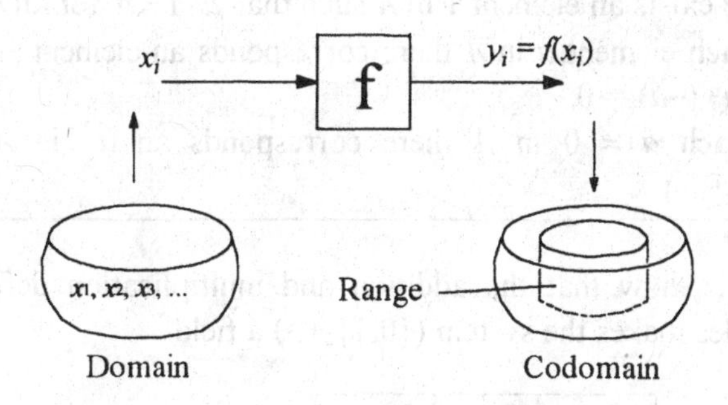

Figure 1.8. The concept of function.

We can think of this in relation to Fig. 1.8 in the following way: Select a time function $v(t)$ and supply it to the black box labeled f. Inside the black box the function $v(t)$ is multiplied by the exponential $e^{-j\omega t}$ and then integrated over all time. The result is the frequency function $V(\omega)$, which pops out of the box and falls into the range. The domain contains all time functions that have a Fourier transform, and the range contains all corresponding frequency functions. The basic idea of a transform is a map where the domain and codomain consist of functions.

Fields

Another necessary concept is that of a *field*. Think of ordinary numbers with the two operations addition and multiplication. This is the model used to define a field. Any set with two operations defined on it that satisfies the following definition is called a field.

Definition 1.2. Let A be a set consisting of the elements $\{a, b, c, \ldots\}$. The *field* $(A, +, \cdot)$ is a set together with two operations, + and $\cdot$. These operations, called *addition* and *multiplication*, together satisfy the following conditions:

(i) $a\cdot(b+c) = a\cdot b + a\cdot c$ for all a, b, c in A.
(ii) $a+(b+c) = (a+b)+c$ for all a, b, c in A.
(iii) $a\cdot(b\cdot c) = (a\cdot b)\cdot c$ for all a, b, c in A.
(iv) $a+b = b+a$ for all a, b in A.
(v) $a\cdot b = b\cdot a$ for all a, b in A.
(vi) There exists an element 0 in A such that $a + 0 = a$ for all a in A.

(vii) There exists an element 1 in A such that $a \cdot 1 = a$ for all a in A.

(viii) To each element a in A there corresponds an element $-a$ in A such that $a+(-a) = 0$.

(ix) To each $a \neq 0$ in A there corresponds an a^{-1} in A such that $a \cdot a^{-1} = 1$.

Example 1.1. Show that the addition and multiplication defined by the following tables makes the system ({0,1},+,·) a field.

$+_2$	0	1
0	0	1
1	1	0

$\cdot_2$	0	1
0	0	0
1	0	1

Solution: This is called the binary field. To show compliance with Definition 1.2, choose values for a, b, and c and determine whether or not properties (i) through (v) are satisfied. For the binary field this gives us 2^3, or 8, possible combinations. All eight combinations must satisfy these properties. (They do.) Then we must find the 0 element and the 1 element (which are 0 and 1) to satisfy properties (vi) and (vii). The additive inverse for 0 is 0 and for 1 it is 1 [property (viii)]. The multiplicative inverse of 1 is 1 because $1 \cdot 1 = 1$. Thus, the addition and multiplication tables above make the system ({0,1},+,·) a field.

■

Example 1.2. Give the tables + and · to make ({0,1,2},+,·) a field.

Solution:

$+_3$	0	1	2
0	0	1	2
1	1	2	0
2	2	0	1

$\cdot_3$	0	1	2
0	0	0	0
1	0	1	2
2	0	2	1

■

The fields used most often in this book are the real number system and the complex number system.

1.3. Power and Energy Signals

A resistor dissipates electrical energy as heat. One of the more memorable experiments in any sophomore circuits lab demonstrates this fact with a candle leaning against a hot resistor. Heat the resistor by impressing a steady voltage of moderate value across it. If the energy dissipated, candle size, and power rating of the resistor are about right, the melting of the candle, which rests against the resistor, will be proportional to the electrical energy dissipated. This experiment demonstrates not only that energy is dissipated as heat, but also that energy is dissipated over time. If the voltage is not too large, the resistor warms to a constant temperature and remains there. In other words, it reaches a steady state. Thus, *average power* is dissipated in the resistor. This implies the following:

1. The heat produced is proportional to average power. Note the emphasis on average power, not instantaneous power.
2. This implies a steady-state signal. Ideally, a steady-state signal exists forever, starting before the world was formed and lasting until long after the world has burned to a crisp. For the lab experiment, however, the signal must last only for the duration of the experiment. Thus, the idealization to infinite time is one of mathematical convenience.

For a continuous-time signal voltage $v(t)$ impressed across a resistor, the average power is defined by

$$P = \lim_{a\to\infty} \frac{1}{2a} \int_{-a}^{a} \frac{v^2(t)}{R}\, dt \tag{1.1}$$

To make this quantity depend only on the signal, set the resistor equal to 1 ohm. The total power, average power, or mean-square value of the signal is defined to be

$$P = \lim_{a\to\infty} \frac{1}{2a} \int_{-a}^{a} v^2(t)\, dt \tag{1.2}$$

The square root of P is the root mean square (rms) value. Notice that current through the resistor could be used just as easily as the voltage in the definition of power. Our definition of power in a signal applies to any signal, whether it is voltage, current, or some nonelectrical function such as pressure or velocity.

This serves to introduce the classification of signals into one of two categories: energy signals or power signals. A signal $v(t)$ is called a *power signal* if the expression in Eq. 1.2 is finite (i.e., $0 < P < \infty$). A signal $v(t)$ is called an *energy signal* if the energy is finite, where the energy E is given by

$$E = \int_{-\infty}^{\infty} v^2(t)\, dt \tag{1.3}$$

This is the total energy, not the instantaneous energy. Our concern is with those signals that can be classified into one of the two categories, but you should understand that there are other signals. A ramp signal, or an increasing exponential signal, has infinite power, and therefore is neither a power nor an energy signal. Our needs do not include such signals, so in this book every signal is either an energy signal or a power signal. Notice that any signal that lasts for only a finite duration is an energy signal. A signal, which lasts for an infinite duration can be either a power or an energy signal, depending on the value of the expressions in Eqs. 1.2 and 1.3.

Why worry about whether a signal is a power or an energy signal? The answer is because the formulas for convolution, correlation, and Fourier transforms are different for each signal class.

Example 1.3. The square pulse in Fig. 1.9 has zero power. To see this, apply Eq. 1.2 with increasing values of a as follows:

for $a = 5$: $$P = \frac{1}{10}\int_0^1 2^2\, dt = \frac{4}{10}$$

for $a = 10$: $$P = \frac{1}{20}\int_0^1 2^2\, dt = \frac{4}{10}$$

for $a = 100$: $$P = \frac{1}{200}\int_0^1 2^2\, dt = \frac{4}{200}$$

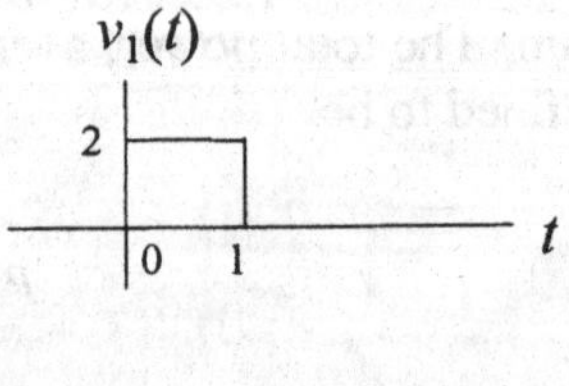

Figure 1.9. Square Pulse.

As a gets larger and larger, the power gets smaller and smaller, until in the limit it is zero. The energy in the pulse is finite, however, because Eq. 1.3 gives

$$E = \int_0^1 2^2 \, dt = 4$$

Since the power is zero and the energy is finite, this is an energy signal. ■

Example 1.4. The periodic square wave in Fig. 1.10 has finite power and infinite energy. Applying Eq. 1.5 gradually, as we did above, gives

for $a = 5$ $$P = \frac{1}{10}\int_{-5}^{5} v_2^2(t)\,dt = \frac{20}{10}$$

for $a = 10$ $$P = \frac{1}{20}\int_{-10}^{10} v_2^2(t)\,dt = \frac{40}{20}$$

for $a = 100$ $$P = \frac{1}{200}\int_{-100}^{100} v_2^2(t)\,dt = \frac{400}{200}$$

Larger and larger values of a result in the same power, 2. Notice that power in a periodic signal of period T is given by the simplified formula

$$P = \frac{1}{T}\int_0^T v^2(t)\,dt \qquad (1.4)$$

which should be familiar from circuit analysis courses. The rms value is the square root of the power. Therefore, the rms value of the signal in this example is $\sqrt{2}$. The energy in the signal is infinite because

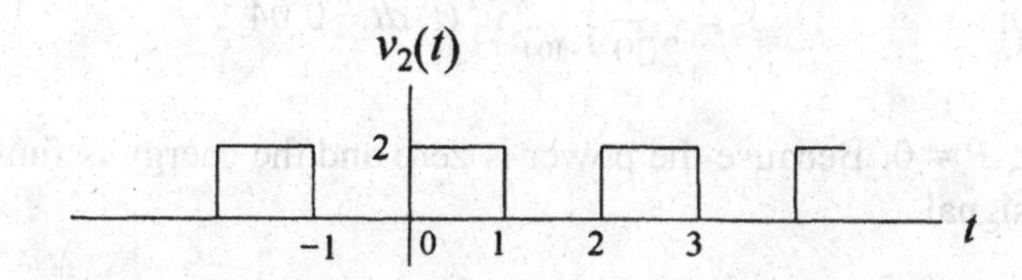

Figure 1.10. Periodic square wave with finite power and infinite energy.

$$E = \int_{-\infty}^{\infty} v_2^2(t)\,dt = \infty$$

Since the energy is infinite and the power is finite, this is a power signal.

■

Any signal that is actually generated and used must be of finite duration, and therefore an energy signal. The classification above applies only to our mathematical model of the signal. We will often say that a signal lasts forever, although this is practically impossible. However, just because a signal lasts forever does not mean that it is a power signal. Here is an example.

Example 1.5. The decaying exponential in Fig. 1.11 lasts forever but has finite energy.

$$E = \int_{-\infty}^{\infty} v_3^2(t)\,dt = 2\int_0^{\infty} 4e^{-2t}\,dt = 4$$

The power is zero, as indicated in the following sequence of integrals.

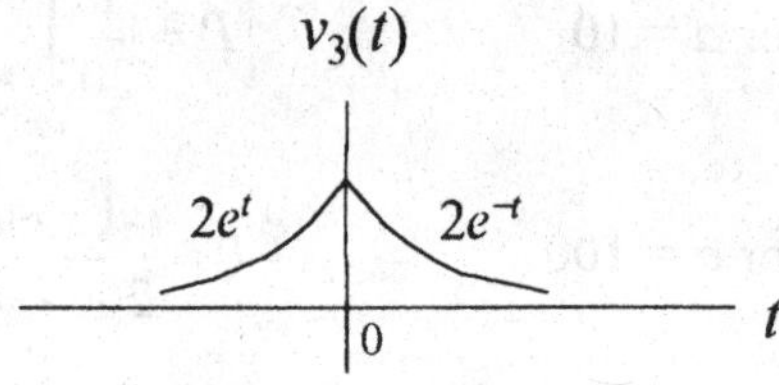

Figure 1.11. Decaying exponential with finite energy.

for $a = 5$: $$P = \frac{1}{10}\int_{-5}^{5} v_3^2(t)\,dt = 0.4$$

for $a = 10$: $$P = \frac{1}{20}\int_{-10}^{10} v_3^2(t)\,dt = 0.2$$

for $a = 100$: $$P = \frac{1}{200}\int_{-100}^{100} v_3^2(t)\,dt = 0.04$$

In the limit, $P = 0$. Because the power is zero and the energy is finite, this is an energy signal.

■

Drill 1.1. Find the power and energy in $v(t) = e^{-2t}u(t)$.

Answer: $P = 0$, $E = ¼$.

For discrete-time signals these definitions become

$$P = \lim_{N \to \infty} \frac{1}{2N+1} \sum_{n=-N}^{N} v^2(n) \tag{1.5}$$

and

$$E = \sum_{n=-\infty}^{\infty} v^2(n) \tag{1.6}$$

In Eq. 1.5, the sum over the range $(-N, N)$ has $2N + 1$ terms in it, so divide by this number of terms to find the average. This is also called the *mean-square value*. The mean-square value is *average power*, although the concept of power loses its meaning for discrete-time signals. The energy in Eq. 1.6 is the *total* energy, not the instantaneous energy.

Example 1.6. The signal $x(n)$ shown in Fig. 1.12 is an energy signal. To see this, apply Eq. 1.5 with increasing values of N to get

for $N = 5$: $$P = \frac{1}{11} \sum_{n=-5}^{5} x^2(n) = \frac{12}{11}$$

for $N = 10$: $$P = \frac{1}{21} \sum_{n=-10}^{10} x^2(n) = \frac{12}{21}$$

for $N = 100$: $$P = \frac{1}{201} \sum_{n=-100}^{100} x^2(n) = \frac{12}{201}$$

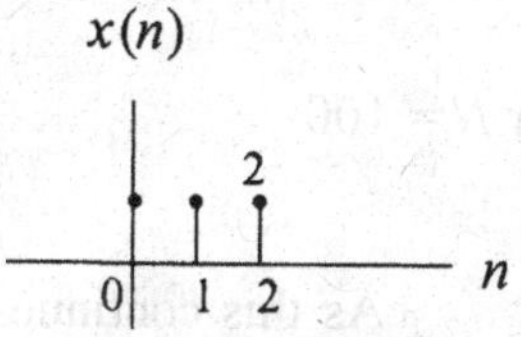

Figure 1.12. Energy signal.

As N gets larger and larger, the power gets smaller and smaller, until in the limit it is zero. The energy in the pulse is finite, however, because Eq. 1.9 gives

$$E = \sum_{n=0}^{2} 2^2 = 12$$

Since the power is zero and the energy is finite, this is an energy signal. ■

Example 1.7 The periodic digital square wave in Fig. 1.13 has finite power and infinite energy. Apply Eq. 1.5 gradually, as above, to get

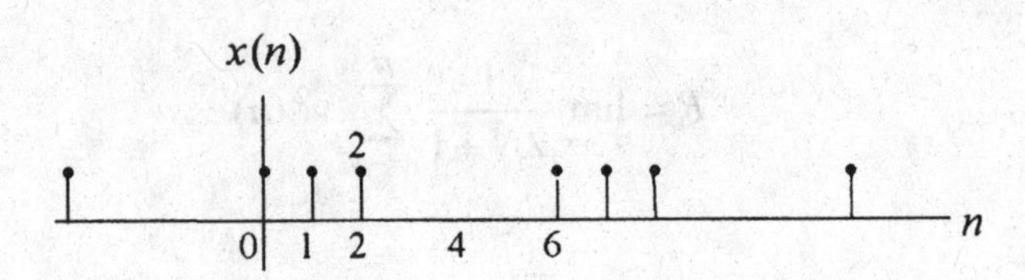

Fig. 1.13. A periodic digital signal with finite power and infinite energy.

for $N = 5$: $$P = \frac{1}{11}\sum_{n=-5}^{5} x^2(n) = \frac{20}{11}$$

for $N = 10$: $$P = \frac{1}{21}\sum_{n=-10}^{10} x^2(n) = \frac{40}{21}$$

for $N = 100$: $$P = \frac{1}{201}\sum_{n=-100}^{100} x^2(n) = \frac{400}{201}$$

As this continues with larger and larger values of N, P approaches the value 2. Notice that here, too, the power in a periodic signal with period N can be found by averaging over only one period.

$$P = \frac{1}{N}\sum_{n=0}^{N-1} x^2(n) \tag{1.7}$$

The rms value is again the square root of the power. The energy is given by

$$E = \sum_{n=-\infty}^{\infty} x^2(n) = \infty$$

Since the energy is infinite and the power is finite, this is a power signal. ■

As with continuous time signals, just because a signal lasts forever does not mean that it is a power signal. Here is an example to illustrate that fact for discrete-time signals.

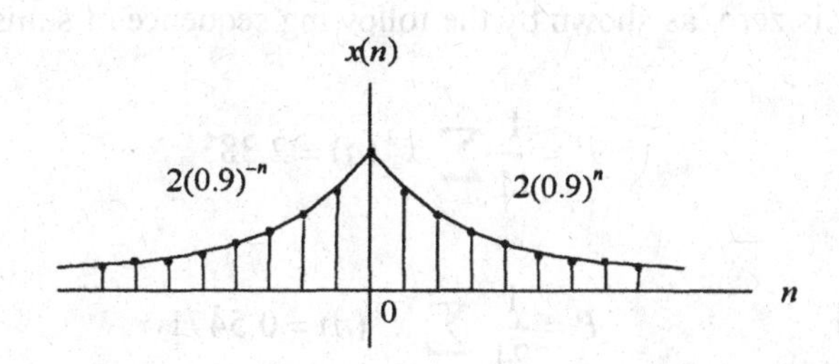

Figure 1.14. Exponential energy signal.

Example 1.8 The exponential signal in Fig. 1.14 lasts forever in both directions, but it is an energy signal. The energy is

$$E = \sum_{n=-\infty}^{\infty} x^2(n) = 2\sum_{n=1}^{\infty} x^2(n) + x^2(0)$$

The terms on the right are derived by noticing that for the waveform in Fig. 1.14, the sum from $-\infty$ to -1 is the same as the sum from 1 to ∞. Then the energy is given by twice the sum from 1 to ∞ plus the term for $n = 0$. The sum $\sum_{n=0}^{\infty}(0.81)^n$ is a geometric series with value given by

$$\sum_{n=0}^{\infty} a^n = \frac{1}{1-a} \qquad |a| < 1 \tag{1.8}$$

The sum in the expression for energy has a lower limit of 1 instead of 0, so calculate the sum using Eq. 1.8 and then subtract the $n = 0$ term. With $a = 0.81$ this gives

$$\sum_{n=1}^{\infty} x^2(n) = 4\sum_{n=0}^{\infty} (0.81)^n - 4 = \frac{4(0.81)}{1-0.81} = 17.0526$$

Therefore, the energy is double this number plus $x^2(0)$.

$$E = 2(17.0526) + 4 = 38.1053$$

The power is zero, as shown by the following sequence of sums.

for $N = 5$: $$P = \frac{1}{11}\sum_{n=-5}^{5} x^2(n) = 2.383$$

for $N = 10$: $$P = \frac{1}{21}\sum_{n=-10}^{10} x^2(n) = 0.5471$$

for $N = 100$: $$P = \frac{1}{201}\sum_{n=-100}^{100} x^2(n) = 0.0552$$

In the limit, $P = 0$. Because the power is zero and the energy is greater than zero but finite, this is an energy signal.

■

We will use only those signals that can be classified into one of the two categories of energy signal or power signal, according to the value of P or E from these definitions. Notice once again that a power signal must last forever, so these definitions apply only to their mathematical model.

Drill 1.2. Find the power and energy in $v(n) = (0.9)^n u(n)$.

Answer: $P = 0$, $E = 5.2632$.

1.4. Deterministic and Random Signals

Different definitions for a random signal exist in the literature, which means that there is no good definition. *Random* means unpredictable, hence a random signal is one with unpredictable values. Wavelet and Fourier transforms apply to random signals as well as to deterministic signals. For now, let us review some history that should help you better understand random noise

The Scottish botanist Robert Brown (1773-1858) was among the leading scientists of his day. He met with and advised the young Charles Darwin (1809-1882) before Darwin's famous voyage on the *Beagle* beginning in 1831. Brown himself had served as the naturalist on a voyage to the still-new and largely unexplored continent of Australia in 1801. His ship returned in 1805 with some 4,000 species of plants.

Brown is remembered particularly for two discoveries. He recognized the small body within cells as a regular feature, and in 1831 gave it the name by which it has been known ever since: *nucleus*, from the Latin word meaning “little nut.” The second discovery had repercussions outside the life sciences. In 1827, he was viewing wheat pollen suspended in water under a microscope when he noticed that the particles were moving about in an irregular fashion. He studied other particles suspended in water, even a particle from the sphinx, and observed this same random motion. This has been called *Brownian motion* ever since. Brown believed that he had discovered a fundamental form of life that was “present in animate and inanimate objects,” and no one proved otherwise during his lifetime.

Ludwig Boltzmann’s tombstone in the central cemetery of Vienna has on it his name, the dates 1844-1906, and the inscription $S = k \log W$. In this formula, S stands for entropy, k is now known as Boltzmann’s constant, and W is a measure of the possible states of nature. This is one of the most remarkable formulas in science, for it quantifies the chaos that exists at the atomic level in matter and applies to any system with random parameters. Boltzmann was an early proponent of atomism at a time when there was considerable debate about the nature of matter. His work influenced the work of Einstein on Brownian motion.

In 1905, Albert Einstein (1879-1955) published five papers involving three developments of major importance in the German journal *Annalen der Physick* (Annals of Physics). His theory of relativity is so startling and contrary to common experience that we often ignore the other two developments. His first paper dealt with the photoelectric effect, and his second paper worked out a mathematical analysis of Brownian motion. Einstein showed that if the water in which the particles were suspended was composed of molecules in random motion according to the kinetic theory of Boltzmann, the result would be the observed Brownian motion of the suspended particles. Einstein went on to say that any system in thermal equilibrium would exhibit this Brownian motion, and he correctly derived the current that flows in a circuit due to the random motion of electrons. Since an electric circuit element such as a resistor is a system in thermal equilibrium, random noise will be present as a voltage across its terminals. Einstein’s formula gives the value of this voltage as

$$V = 4kTR \times \mathrm{Bw}$$

where V is the rms voltage, k is Boltzmann’s constant, T is the temperature in Kelvin, R is the value of the resistance in ohms, and Bw is the bandwidth of the measuring instrument. For example, a 1-kΩ resistor at room temperature, measured by an ordinary voltmeter, has an rms voltage across its terminals of 3 to 6 μV. The waveform in Fig. 1.15 shows a typical oscilloscope image of this voltage.

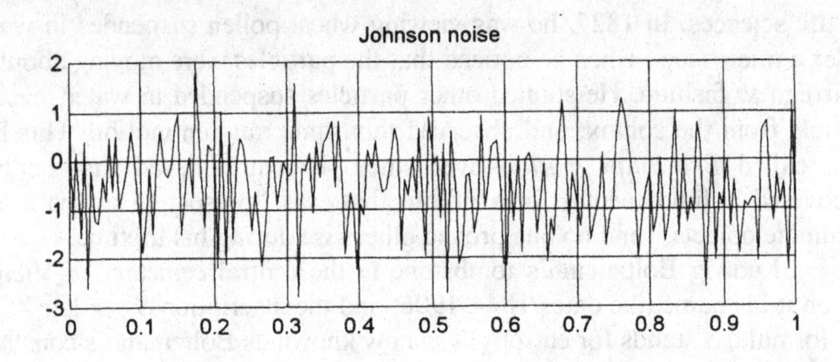

Figure 1.15. Typical Johnson noise.

This noise, called *Johnson noise* after the scientist who first measured it at Bell Labs in 1928, limits communication. A distant transmitter may be so weak that this noise obliterates the signal. One way to combat this limitation is to reduce the temperature in the front end of the receiver. Thus, some sensitive receivers operate in a cold environment, with the components in the first few stages of amplification reduced in temperature to near absolute zero.

1.5. Fourier and Haar Transforms

Signals fall naturally into one of four categories:

1. Continuous-time energy signals.
2. Continuous-time power signals.
3. Discrete-time energy signals.
4. Discrete-time power signals.

There is one form of the Fourier transform for each category

1. Continuous-time Fourier transform (CTFT).
2. Continuous-time Fourier series (CTFS).
3. Discrete-time Fourier transform (DTFT).
4. Discrete-time Fourier series (DTFS).

The matchup goes like this:

1. Continuous-time energy signals. ↔ CTFT.
2. Continuous-time power signals. ↔ CTFS.
3. Discrete-time energy signals. ↔ DTFT.
4. Discrete-time power signals. ↔ DTFS.

The fast Fourier transform is a fast version of the DTFS and does not apply to any of the other transforms. The DTFS is defined by

$$V(k) = \sum_{n=0}^{N-1} v(n) e^{-j2\pi nk/N} \tag{1.9}$$

This is the only one of the four Fourier transforms that will be of much concern to us in this book.

Note that Eq. 1.9 is really N equations, one for each value of k.

$$V(0) = \sum_{n=0}^{N-1} v(n) e^{0} = \sum_{n=0}^{N-1} v(n)$$

$$V(1) = \sum_{n=0}^{N-1} v(n) e^{-j2\pi n/N}$$

$$\vdots \qquad\qquad \vdots$$

$$V(N-1) = \sum_{n=0}^{N-1} v(n) e^{-j2\pi n(N-1)/N}$$

There is a different exponential term used in each equation, starting with e^0, $e^{-j2\pi n/N}$, and progressing to $e^{-j2\pi n(N-1)/N}$. These are the basis functions for the expansion. (Chapter 3 covers the concept of basis functions in detail.) In other words, the DTFS is an expansion of $v(n)$ in terms of these exponentials. These exponential terms are usually written in a standard form. Let

$$W_N = e^{-j2\pi/N} \tag{1.10}$$

Then

$$W_N^0 = e^0$$

$$W_N^1 = e^{-j2\pi/N}$$

$$W_N^2 = e^{-j2\pi 2/N}$$

$$\vdots \qquad \vdots$$

$$W_N^{N-1} = e^{-j2\pi(N-1)/N}$$

These N equations can be put in matrix form. For example, if $N = 4$, the matrix is given by

$$\begin{bmatrix} V(0) \\ V(1) \\ V(2) \\ V(3) \end{bmatrix} = \begin{bmatrix} W_4^0 & W_4^0 & W_4^0 & W_4^0 \\ W_4^0 & W_4^1 & W_4^2 & W_4^3 \\ W_4^0 & W_4^2 & W_4^4 & W_4^6 \\ W_4^0 & W_4^3 & W_4^6 & W_4^9 \end{bmatrix} \begin{bmatrix} v(0) \\ v(1) \\ v(2) \\ v(3) \end{bmatrix} \tag{1.11}$$

Using this notation, Eq. 1.9 can be written as

$$V(k) = \sum_{n=0}^{N-1} v(n) W_N^n \tag{1.12}$$

Example 1.9. Find the DTFS for $v(n) = \{1, 2, 3, 4\}$.

Solution: Here, $N = 4$, so we can use Eq. 1.11 directly. Multiplying by the matrix in Eq. 1.11 gives

$$\begin{bmatrix} V(0) \\ V(1) \\ V(2) \\ V(3) \end{bmatrix} = \begin{bmatrix} W_4^0 & W_4^0 & W_4^0 & W_4^0 \\ W_4^0 & W_4^1 & W_4^2 & W_4^3 \\ W_4^0 & W_4^2 & W_4^4 & W_4^6 \\ W_4^0 & W_4^3 & W_4^6 & W_4^9 \end{bmatrix} \begin{bmatrix} v(0) \\ v(1) \\ v(2) \\ v(3) \end{bmatrix}$$

$$
= \begin{bmatrix} 1 & 1 & 1 & 1 \\ 1 & -j & -1 & j \\ 1 & -1 & 1 & -1 \\ 1 & j & -1 & -j \end{bmatrix} \begin{bmatrix} 1 \\ 2 \\ 3 \\ 4 \end{bmatrix} = \begin{bmatrix} 10 \\ -2+j2 \\ -2 \\ -2-j2 \end{bmatrix}
$$

■

Haar Transform

The Haar transform is one (of several) wavelet transforms that can be calculated with a formula. The formula (actually, two formulas) is given by

$$c_{00} = \int_0^1 v(t)\,\varphi_{00}(t)\,dt \tag{1.13a}$$

$$d_{kj} = \int_0^1 v(t)\,\psi_{kj}(t)\,dt \tag{1.13b}$$

where the basis functions φ_{00} and ψ_{kj} are as shown in Fig. 1.16. The function $\varphi_{00}(t)$ is called the *scaling function*, and the $\psi_{kj}(t)$ are called *wavelets*.

This diagram shows the level 0, 1, and 2 Haar basis functions, but these can continue to higher and higher levels. Level 0 contains the functions $\varphi_{00}(t)$ (called the *scaling function*) and $\psi_{00}(t)$ (called the *mother wavelet*). Level 1 contains the two functions $\psi_{10}(t)$ and $\psi_{11}(t)$. Level 2 contains the four functions $\psi_{20}(t)$, $\psi_{21}(t)$, $\psi_{22}(t)$, and $\psi_{23}(t)$. Each higher level contains twice as many wavelets $\{\psi_{jk}\}$. Thus, level 3 contains eight wavelets, each just half as long as the level 2 wavelets and scaled appropriately to have unit energy. To calculate the Haar transform, expand the time function in terms of these basis functions. Equation 1.13 gives the formulas for calculating the transform.

Example 1.10. Calculate the Haar transform of the signal in Fig. 1.17.

Solution: Applying Eq. 1.13 to the function v(t) gives the following numbers.

$$c_{00} = \int_0^1 v(t)\,\varphi_{00}(t)\,dt = 0$$

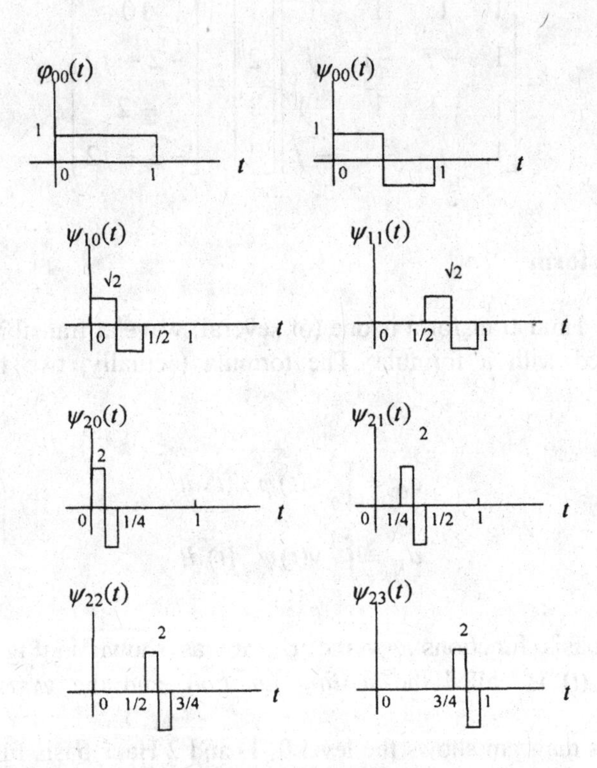

Figure 1.16. Haar basis functions to level 2.

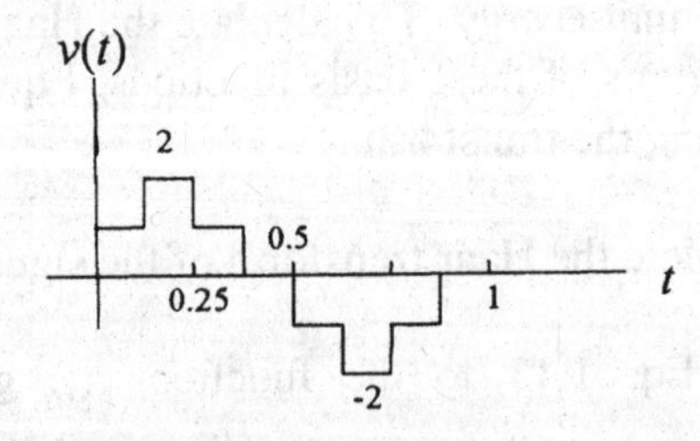

Figure 1.17. The function $v(t)$.

$$d_{00} = \int_0^1 v(t)\psi_{00}(t)\,dt = 1$$

$$d_{10} = \int_0^{0.5} v(t)\psi_{10}(t)\,dt = \frac{\sqrt{2}}{4}$$

$$d_{11} = \int_{0.5}^{1} v(t)\psi_{11}(t)\,dt = -\frac{\sqrt{2}}{4}$$

$$d_{20} = \int_0^{0.25} v(t)\psi_{20}(t)\,dt = -\frac{1}{4}$$

$$d_{21} = \int_{0.25}^{0.5} v(t)\psi_{21}(t)\,dt = \frac{1}{4}$$

$$d_{22} = \int_{0.5}^{0.75} v(t)\psi_{22}(t)\,dt = \frac{1}{4}$$

$$d_{23} = \int_{0.75}^{1} v(t)\psi_{23}(t)\,dt = -\frac{1}{4}$$

Thus, the first eight terms of the Haar transform of $v(t)$ consist of the eight values $\left\{0,\ 1,\ \frac{\sqrt{2}}{4},\ -\frac{\sqrt{2}}{4},\ -\frac{1}{4},\ \frac{1}{4},\ \frac{1}{4},\ -\frac{1}{4}\right\}$.

■

Equation 8.12 gives the equation for the inverse wavelet transform. When applied to the function in Fig. 1.17, it takes the form

$$v(t) = c_{00}\varphi_{00}(t) + d_{00}\psi_{00}(t) + d_{10}\psi_{10}(t) + \cdots + d_{23}\psi_{23}(t)$$

Example 1.11. Show that the inverse transform gives the function $v(t)$ in Fig. 1.17.

Solution: Multiplying the functions in Fig. 1.16 by the coefficients, $c_{00} = 0$, $d_{00} = 1$, and so on, results in the functions in Fig. 1.18. If these functions are summed, the result is $v(t)$ in Fig. 1.17. (Try it if you don't believe it.)

■

From these examples one can see many similarities between Fourier and wavelet transforms. Time functions are expanded in terms of basis functions in both transforms. These basis functions are exponentials for the Fourier transform, and they are square pulses for the Haar transform. The coefficients are calculated by an inner product operation [i.e., by integrating the product of $v(t)$ and basis function].

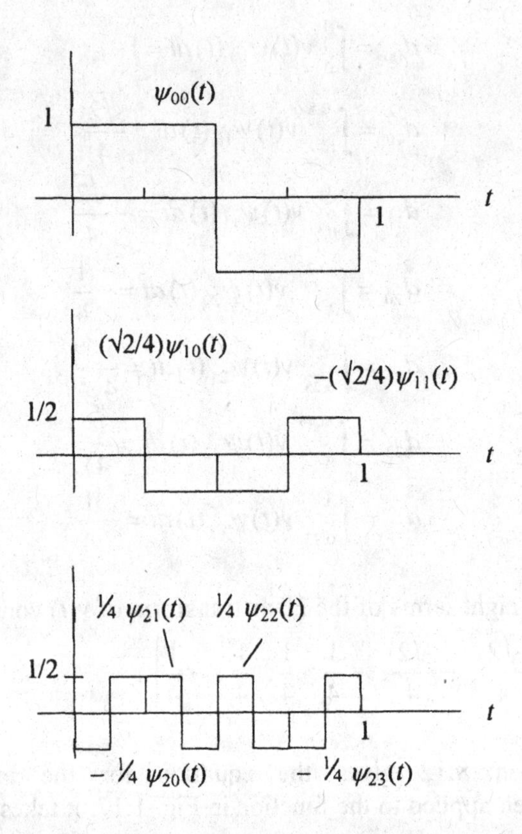

Figure 1.18. Sum these functions to obtain $v(t)$.

However, there are differences that are not apparent in these simple examples. There are many wavelet basis functions other than the Haar functions, and there is one wavelet transform for each set of basis functions. This is similar to the Fourier transforms, where there are four forms of the Fourier transform, but not similar in that there is a vast array of wavelet basis functions. Another important difference is in the way the coefficients for the wavelet transform are calculated. Note that in the above example using an inner product formulation, the formula for the basis function must be known. Most wavelet coefficients are calculated in a different way, using multirate sampling theory. There it is necessary to know only the filter coefficients. This method does not use the basis functions. We will devote much of this book to this procedure

Chapter 2
Vectors

Ask an engineer to define vector, and the answer will probably be, "a directed magnitude." Geometric vectors are directed magnitudes, but there are other kinds of vectors besides geometric vectors, and these other vectors don't have magnitude and direction. Vectors, like anything else, are defined by listing properties. The important properties of geometric vectors are listed in the definition below, and anything else that satisfies these properties is called a *set* of vectors. It turns out that many other things satisfy this definition, including waveforms. Many important signal properties are actually vector properties. It is often said that signal processing is geometry.

The concept of "vector" applies to much more than geometric vectors. It applies to such diverse things as waveforms, matrices, and numbers. In fact, a theorem makes it clear that the set of all functions defined on some set A whose codomain is a *vector space* is itself a vector space. Since the real number system R is a vector space, the set of all mappings from A to R is a vector space. This means that signals form a vector space, and that many other sets are vector spaces. For example, a random variable is a function whose domain is the sample space (the set of all possible experimental outcomes) with codomain R. Therefore, the set of all random variables that could be defined on the sample space is a vector space. This explains the term *orthogonality principle* in mean square estimation. Since random variables are vectors, geometry applies to them.

A vector space without additional properties is practically useless. The addition of an inner product makes it possible to do geometry, the operations performed in signal processing. An inner product induces a norm, which in turn induces a distance measure. These are the operations necessary for digital signal processing. The addition of calculus allows us to do analog signal processing. A *Banach space* is a set that contains all its limit points, making calculus possible. The combination of an inner product space with a Banach space makes it possible to do both geometry and calculus. This combination is called a *Hilbert space*, the space necessary for analog signal processing.

John von Neumann (1903-1957) originated several new concepts. Not only did he originate Hilbert space but he also developed cellular automata, game theory, and invented the stored program concept for digital computers. All four of these concepts are in common use today. Mathematicians and scientists use Hilbert spaces, economists use game theory, robots are commonplace, and the stored program concept makes possible the personal computer. One could argue that von Neumann influenced society more than any other single person did in the past 100 years, for he made possible the information age.

Chapter Goals: After completing this chapter, you should be able to do the following:

- Define *vector space*, including all the properties.
- Define *metric space*, *norm*, and *inner product*.
- Find the norm and metric induced by a given inner product.
- Write down the usual inner product for each of the four signal types, as given in Eqs. 2.26 through 2.29. Then, from this, you should be able to derive the usual norm and metric for each signal type.

2.1. Vector Space

Most people view vectors in the same manner as the three men who looked through long tubes at an elephant. The first man said an elephant was like a snake, because he saw the tail. The second man said an elephant was like a tree, because he saw a leg. The third man said an elephant was like a barn, because he saw the side. Most people look at vectors through the tube that sees an arrow ⟶ (i.e., a "directed magnitude"). They never realize that there are other equally valid views.

Once a person starts thinking in a certain direction, it is almost impossible to change his or her thinking. Yet that is what one must do in order to benefit from the concept of vector in signal processing. Here is a puzzle that demonstrates the difficulty in changing one's thinking.

Puzzle: Figure 2.1 shows the top and front views of a common object. Draw the side view.

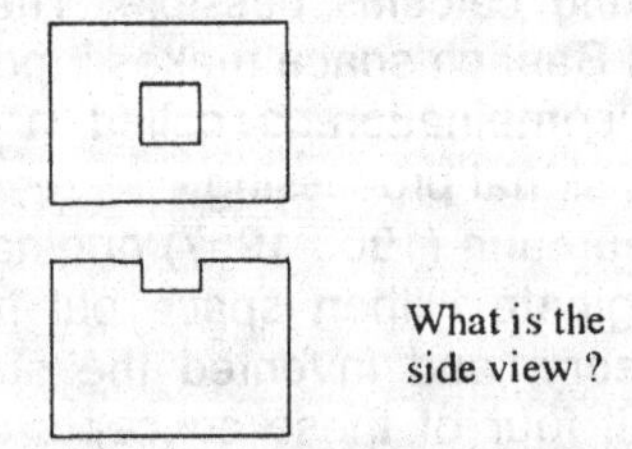

Figure 2.1. The objective of this puzzle is to draw the side view.

There are no lines missing, none of the lines should be dotted, and the puzzle is not impossible. See if you can solve it before peeking at the answer at the end of the chapter.

You will probably have no difficulty, because you have been told that you must change your thinking in order to solve the puzzle. Be prepared to change your thinking on vectors. If you are typical, it will not be easy. The concept of vector encompasses much more than geometric vectors (pointy things), and this seems to be an especially difficult concept to realize.

The definition of vector applies to a set, not just one element. This set collectively satisfies a list of properties. Several different but equivalent definitions exist in the literature. One definition differs from another in the way they choose to present the defining properties. The following definition makes clear that two operations are involved, along with seven other properties.

Definition 2.1. A *vector space* is a set $V = \{v_i\}$ together with a field of scalars $A = \{a_i\}$ that has the following two operations and seven properties.

1. Two vectors can add together to obtain a third vector. This mechanism combines two vectors to produce a third.
2. A scalar can multiply a vector to obtain another vector. This mechanism combines a scalar with a vector to produce another vector.

Using these two operations, vector addition and scalar multiplication, the following properties must hold for all $v_i \in V$ and all $a_i \in A$.

(1) $v_1 + v_2 = v_2 + v_1$.
(2) $(v_1 + v_2) + v_3 = v_1 + (v_2 + v_3)$.
(3) $a_1(v_1 + v_2) = a_1v_1 + a_1v_2$.
(4) $(a_1 + a_2)v_i = a_1v_i + a_2v_i$.
(5) $(a_1 a_2)v_i = a_1 (a_2v_i)$.
(6) $1 \cdot v_i = v_i$.
(7) There exists a unique vector v_0, called the *zero vector*, such that for all vectors v_i we have $0 \cdot v_i = v_0$.

Recall that a field is modeled after the real number system with the two operations addition and multiplication. In this book we use both the real field and the complex field, so for *scalar* you can substitute number, where this means either a real number or a complex number. (Since real numbers are a special case of complex numbers, it is proper to view all scalars as complex numbers.)

The properties above define a vector space. Notice that this definition does not apply to a single vector. Instead, it applies to a set. A vector space is a set together with two functions, a mapping $V \times V \to V$ that defines vector addition, and a mapping $A \times V \to V$ that defines scalar multiplication. (The notations $V \times V$ and $A \times V$ stand for the set cross product.) Any set of objects with these two functions that satisfies the seven properties is a vector space. Furthermore, these vectors are as legitimate as geometric vectors, even though they may not have magnitude and direction.

Ordinary geometric vectors satisfy these properties and therefore form a vector space. To show this, let v_1, v_2, ... be a set of ordinary geometric vectors. These are the familiar directed magnitudes of geometry. Let a_1, a_2, ... be ordinary numbers (from the real number system). These numbers are called scalars but they are just ordinary numbers. Two vectors add together in any order [property (1)]. One can sum three vectors by first adding $v_1 + v_2$ and then adding v_3, or by adding $v_2 + v_3$ before adding v_1, to get the same result either way [property (2)]. Continuing in this way, you can see that geometric vectors satisfy each property listed.

However, geometric vectors are not the only objects that satisfy Definition 2.1. All 3×2 matrices satisfy these properties. We don't usually call a matrix a vector, but all matrices of the same dimension form a vector space. It is customary to label $n \times 1$ matrices (column vectors) as vectors, but because their components are so closely associated with the Euclidean components of geometric vectors, people often think of them as geometric vectors instead of as matrices.

The set of real numbers satisfies these properties. Here the numbers play a dual role: They are both vector and scalar.

Our interest in vectors stems from the fact that all waveforms that can be generated in the laboratory form a vector space. Here instead of v_1, v_2, ..., we have $v_1(t)$, $v_2(t)$, ... Waveforms can add together to form another waveform. Scalars can multiply waveforms to change the amplitude. And there is a zero vector, namely, $v_0(t) = 0$, for all time, which satisfies property (7). (These same statements apply equally to a set of discrete-time waveforms.) Therefore, you can see that waveforms (signals) are vectors.

Here is where your thinking must change. A waveform is a vector in the same sense that a directed magnitude is a vector. To say that one can "represent" a waveform by a vector shows a lack of understanding. There is no need to represent a waveform by a vector since a waveform *is* a vector.

Example 2.1. Consider the set of all triangular waveforms $v(t) = at$, $0 < t < 1$. Figure 2.2 shows two of the infinite number of waveforms in this set. Which of the following is a vector space? That is, when can we call these waveforms vectors?
(a) a is any real number in the interval $-1 < a < 1$.
(b) a is any complex number whatsoever.
(c) a is any real number whatsoever.

Solution: (a) This set does not satisfy the addition operation. That is, if

$$v_1(t) = 0.4t \quad \text{and} \quad v_2(t) = 0.8t$$

then $v_1(t) + v_2(t) = 1.2t$, which is too big to be in the set. Therefore, adding two waveforms together does not always produce another waveform that is in the set. Neither does this set satisfy the "multiplication by scalars" property.

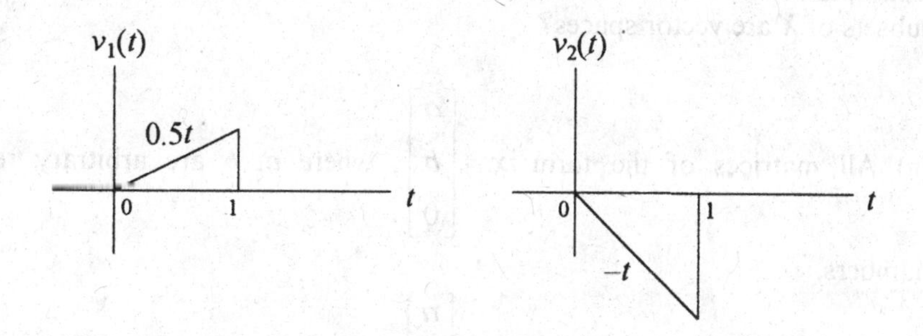

Figure 2.2. Triangular waveforms.

(b) This set satisfies the vector addition and scalar multiplication operations in Definition 2.1. It also satisfies each of the seven properties. The zero vector is the waveform $v(t) = 0$, $0 < t < 1$. Therefore, this is a vector space.
(c) This set also satisfies all properties. In fact, this is a subspace of b). A subspace is a subset that satisfies all properties of a vector space. Notice that the set in (a) is also a subset of (b), but it is not a subspace.

■

Example 2.2. Let X be the set of all possible waveforms. Which of the following subsets of X are vector spaces?
(a) A particular waveform generator can produce waveforms that are square, sinusoidal, triangular, and random. Does this subset form a vector space?
(b) Does the set of all square waveforms that this generator can produce form a vector space?

Solution: (a) Adding a square wave and a sinusoid will produce a waveform that is not in the subspace. That is, the sum will not be a square, sinusoidal, triangular, or random signal. Therefore, this is not a vector space.
(b) Here, care must be taken. Make the period and the starting time (phase) of all square waves the same. Then these square waves "almost" form a vector space. The trouble is that the maximum output amplitude is fixed, giving the same problem as in Example 2.1a. If the waveform generator could produce waveforms of arbitrary amplitude, this would be a vector space.

■

Example 2.3. Let X be the set of all 3×1 matrices. Which of the following subsets of X are vector spaces?

(a) All matrices of the form $x = \begin{bmatrix} a \\ b \\ 0 \end{bmatrix}$, where a, b are arbitrary real numbers.

(b) All matrices of the form $x = \begin{bmatrix} a \\ b \\ 2 \end{bmatrix}$, where a, b are arbitrary real numbers.

(c) All matrices x such that $Ax = \begin{bmatrix} 0 \\ 0 \\ 0 \end{bmatrix}$, where A is a 3×3 matrix.

Solution: (a) This is a vector space. Add two of these matrices to get

$$\begin{bmatrix} a_1 \\ b_1 \\ 0 \end{bmatrix} + \begin{bmatrix} a_2 \\ b_2 \\ 0 \end{bmatrix} = \begin{bmatrix} a_3 \\ b_3 \\ 0 \end{bmatrix}$$

which is another member of this set. In addition, multiplying by a scalar produces another member of this set. All seven properties are satisfied, so this is a vector space.
(b) This is not a vector space. Pick any member of the set and multiply by 2. This produces a 3×1 matrix that is not a member of the set because the last component is 4, not 2.
c) This is a vector space. In fact, it is called the null space of the matrix A. Adding two vectors gives

$$Ax_1 + Ax_2 = \begin{bmatrix} 0 \\ 0 \\ 0 \end{bmatrix} + \begin{bmatrix} 0 \\ 0 \\ 0 \end{bmatrix} = \begin{bmatrix} 0 \\ 0 \\ 0 \end{bmatrix}$$

In addition, multiplying by any scalar produces the zero vector. All seven properties are satisfied, so this is a vector space. ■

2.2 Metric Space

Definition 2.1 imposes little structure on vector spaces. Multiplication by scalars and addition of vectors are the only operations defined. Nothing has been said about how to measure the length of a vector, how to determine the distance between two vectors, or how to find the dot product of two vectors. Doing geometry requires all three of these things. A vector space in which we can do geometry is called an *inner product* (or *dot product*) space. Once the dot product has been defined, a measure of length and distance is imposed automatically. This means that if a dot product has been defined for a vector space, we can do geometry because a measure of length and distance is imposed on the space automatically by the dot product. We begin with the concept of distance.

Definition 2.2. A *metric space* (X,d) is a set X with a function $d\, X \times X \to R$ which assigns to every pair of elements a and b of X a number $d(a,b) \geq 0$ which has the properties
(i) $d(a,b) = 0$ iff $a = b$.

(ii) $d(a,b) = d(b,a)$ (symmetry)
(iii) $d(a,b) \le d(a,c) + d(c,b)$ (triangle inequality)

The number $d(a,b)$ is the distance between a and b, and d is called a *metric* or *distance function*.

For each set, there are usually several ways to measure distance. For example, the usual metric for 3×1 matrices is defined by

$$d_2(x_1, x_2) = d_2\left(\begin{bmatrix} a_1 \\ b_1 \\ c_1 \end{bmatrix}, \begin{bmatrix} a_2 \\ b_2 \\ c_2 \end{bmatrix}\right)$$

$$= \sqrt{|a_1 - a_2|^2 + |b_1 - b_2|^2 + |c_1 - c_2|^2} \tag{2.1}$$

However, there are other ways to measure distance in this space. Another valid metric is given by

$$d_1(x_1, x_2) = |a_1 - a_2| + |b_1 - b_2| + |c_1 - c_2| \tag{2.2}$$

The function d_1 satisfies all three properties of Definition 2.2. Another valid metric is given by

$$d_p(x_1, x_2) = \left[\, |a_1 - a_2|^p + |d_1 - d_2|^p + |c_1 - c_2|^p \right]^{1/p} \tag{2.3}$$

where p is an integer, $p \ge 1$.

Example 2.4. For any nonempty set X, define

$$d(a,b) = \begin{cases} 1 \text{ if } b \ne a \\ 0 \text{ if } b = a \end{cases}$$

(a) Determine if this is a valid metric.
(b) Let $X = R$. Find all distances between $a = 1$, $b = 1.5$, and $c = 1.5$.

Solution: (a) Clearly this d, called the *trivial* or *discrete metric*, satisfies (i), (ii) and (iii) of Definition 2.2.

b) $d(a,b) = 1$, $d(a,c) = 1$, $d(b,c) = 0$. [Note that $d(a,b) \le d(a,c) + d(b,c)$.]

■

Example 2.5. Our interest is in waveforms. Four metrics for the set of all continuous-time waveforms $\{v_i(t)\}$ defined over $a \le t \le b$ are given by

$$d_\infty(v_1, v_2) = \max_{a \le t \le b} |v_1(t) - v_2(t)| \tag{2.4}$$

$$d_1(v_1, v_2) = \int_a^b |v_1(t) - v_2(t)|\, dt \tag{2.5}$$

$$d_2(v_1, v_2) = \left[\int_a^b |v_1(t) - v_2(t)|^2\, dt\right]^{1/2} \tag{2.6}$$

$$d_p(v_1, v_2) = \left[\int_a^b |v_1(t) - v_2(t)|^p\, dt\right]^{1/p} \tag{2.7}$$

Find the distance between $v_1(t) = 0.5t$ and $v_2(t) = -t$, defined over the interval $0 \le t \le 1$ for each metric. Let $p = 3$ in the last metric.

Solution: Figure 2.3 shows the two waveforms, along with $| v_1(t) - v_2(t) |$. The maximum value of $| v_1(t) - v_2(t) |$ is 1.5, so

$$d_\infty(v_1, v_2) = 1.5$$

Also,

$$d_1(v_1, v_2) = \int_0^1 |v_1 - v_2|\, dt = \int_0^1 1.5t\, dt = 0.75$$

$$d_2(v_1, v_2) = \left[\int_0^1 |v_1 - v_2|^2\, dt\right]^{1/2} = \left[\int_0^1 \frac{9}{4} t^2\, dt\right]^{1/2} = \left(\frac{3}{4}\right)^{1/2}$$

$$d_3(v_1, v_2) = \left[\int_0^1 |v_1 - v_2|^3\, dt\right]^{1/3} = \left[\int_0^1 \frac{27}{8} t^3\, dt\right]^{1/3} = \left(\frac{27}{32}\right)^{1/3}$$

■

Drill 2.1. Let $v_1(t) = 0.5t$, $0 \le t < 1$ (as in Fig. 2.3), and let $v_2(t) = u(t) - u(t-1)$ (a square pulse). Find the distance between these two vectors for each metric in Eqs. 2.4 through 2.7. Use $p = 3$ in Eq. 2.7.

Answer: $d_\infty = 1, \quad d_1 = 3/4, \quad d_2 = \sqrt{7/12}, \quad d_3 = \sqrt[3]{15/32}$

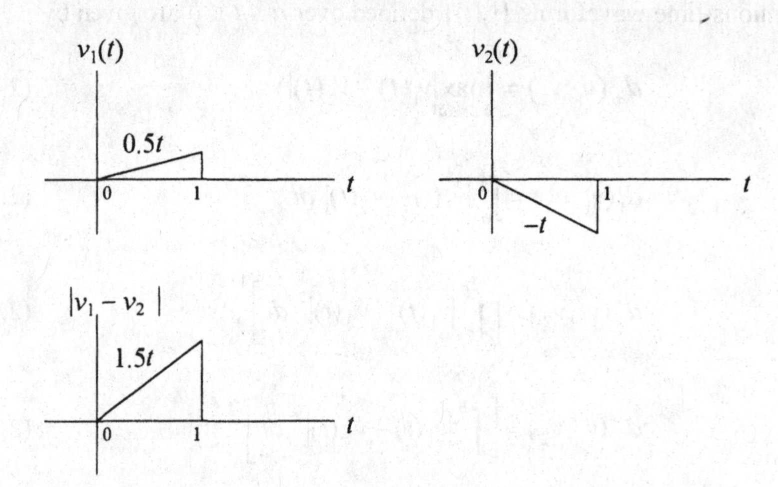

Figure 2.3. Waveforms for Example 2.5.

Example 2.6. For discrete-time waveforms, the metrics corresponding to those in Example 2.5 are given by

$$d_\infty(v_1, v_2) = \max_{0 \le n \le N-1} |v_1(n) - v_2(n)| \tag{2.8}$$

$$d_1(v_1, v_2) = \sum_{n=0}^{N-1} |v_1(n) - v_2(n)| \tag{2.9}$$

$$d_2(v_1, v_2) = \left[\sum_{n=0}^{N-1} |v_1(n) - v_2(n)|^2 \right]^{1/2} \tag{2.10}$$

$$d_p(v_1, v_2) = \left[\sum_{n=0}^{N-1} |v_1(n) - v_2(n)|^p \right]^{1/p} \tag{2.11}$$

Find the distance between $v_1(n) = 4$ and $v_2(n) = n + 1$, where $0 \le n \le 3$. Use $p = 3$ in the last metric.

Solution: Figure 2.4 shows these waveforms along with $|v_1(n) - v_2(n)|$ and the square of this magnitude. The maximum value of $|v_1(n) - v_2(n)|$ is 3, so

$$d_\infty(v_1, v_2) = 3$$

Also,

$$d_1(v_1, v_2) = \sum_{n=0}^{N-1} |v_1(n) - v_2(n)| = 3 + 2 + 1 = 6$$

$$d_2(v_1, v_2) = \left[\sum_{n=0}^{N-1} |v_1(n) - v_2(n)|^2\right]^{1/2} = (9 + 4 + 1)^{1/2} = \sqrt{14}$$

$$d_p(v_1, v_2) = \left[\sum_{n=0}^{N-1} |v_1(n) - v_2(n)|^p\right]^{1/p} = (27 + 8 + 1)^{1/3} = \sqrt[3]{36}$$

■

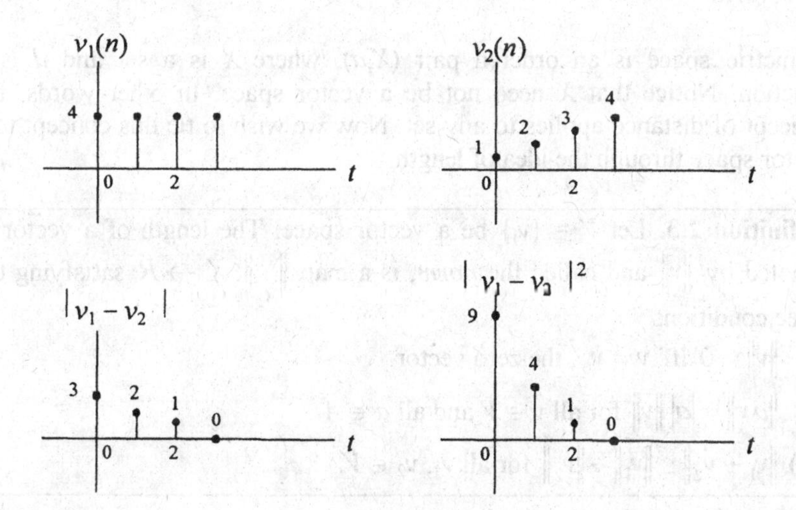

Figure 2.4. Discrete-time waveforms for Example 2.6.

These examples illustrate that different ways to measure distance result in different values for distance between the same vectors. This is not an academic exercise without application. Different measures for distance apply to different applications. For example, the most used distance measure in coding theory is d_1, and both d_2 and d_∞ find use in system theory. Note that the waveforms in system theory are vectors, in every way equivalent to geometric vectors, and in many ways more important. *System theory* is often called *applied geometry*, with good reason.

Drill 2.2. Find the distance between the two vectors in Fig. 2.5 for each metric in Eqs. 2.8 through 2.11. Let $p = 3$ in Eq. 2.11.

Answer:

$$d_\infty = 2, \quad d_1 = 4, \quad d_2 = \sqrt{8}, \quad d_3 = \sqrt[3]{16}.$$

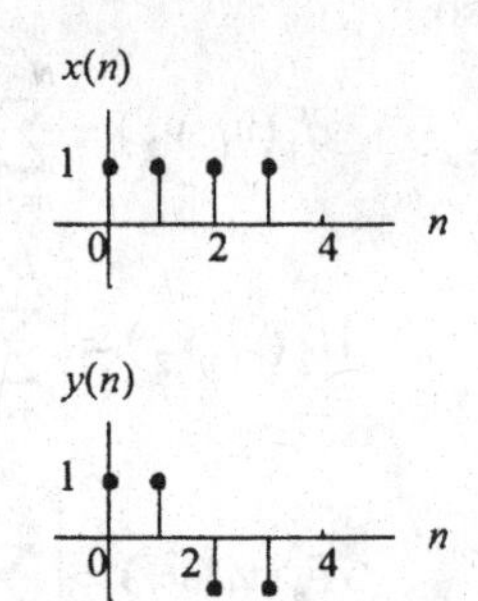

Figure 2.5. Two vectors.

2.3. Norm

A metric space is an ordered pair (X,d), where X is a set and d is a function. Notice that X need not be a vector space. In other words, the concept of distance applies to any set. Now we wish to tie this concept to a vector space through the idea of length.

Definition 2.3. Let $V = \{v_i\}$ be a vector space. The length of a vector v, denoted by $\|v\|$ and called the *norm*, is a map $\|\cdot\|: X \to R$ satisfying the three conditions

(i) $\|v\| = 0$ iff $v = v_0$, the zero vector.

(ii) $\|av\| = |a|\,\|v\|$ for all $v \in V$ and all $a \in A$.

(iii) $\|v_1 + v_2\| \le \|v_1\| + \|v_2\|$ for all $v_1, v_2 \in V$.

The norm of a vector is a measure of the size or length of the vector, and properties (ii) and (iii) remind us that we can define norms on sets V where addition $v_1 + v_2$ and scalar multiplication av make sense. For that reason, V must be a vector space.

Example 2.7. Four norms for the set of all continuous-time waveforms $\{v_i(t)\}$ defined over $a \le t \le b$ are given by

$$\|v\|_\infty = \max_{a \le t \le b} |v(t)| \tag{2.12}$$

$$\|v\|_1 = \int_a^b |v(t)|\, dt \tag{2.13}$$

$$\|v\|_2 = \left[\int_a^b |v(t)|^2\, dt\right]^{\frac{1}{2}} \tag{2.14}$$

$$\|v\|_p = \left[\int_a^b |v(t)|^p\, dt\right]^{\frac{1}{p}} \tag{2.15}$$

Find the length of $v_1(t) = 0.5t$ and $v_2(t) = -t$, defined over the interval $0 \le t \le 1$ for each norm. Let $p = 3$ in the last norm. (Figure 2.2 shows these waveforms.)

Solution:

$$\|v_1\|_\infty = 0.5 \qquad \|v_2\|_\infty = 1$$

$$\|v_1\|_1 = \int_0^1 0.5t\, dt = \frac{1}{4}t^2\Big|_0^1 = \frac{1}{4} \qquad \|v_2\|_1 = \int_0^1 t\, dt = \frac{1}{2}t^2\Big|_0^1 = \frac{1}{2}$$

$$\|v_1\|_2 = \left[\int_0^1 \frac{1}{4}t^2\, dt\right]^{1/2} = \left(\frac{1}{12}\right)^{1/2} \qquad \|v_2\|_2 = \left[\int_0^1 t^2\, dt\right]^{1/2} = \left(\frac{1}{3}\right)^{1/2}$$

$$\|v_1\|_3 = \left[\int_0^1 \frac{1}{8}t^3\, dt\right]^{1/3} = \left(\frac{1}{32}\right)^{1/3} \qquad \|v_2\|_3 = \left[\int_0^1 t^3\, dt\right]^{1/3} = \left(\frac{1}{4}\right)^{1/3}$$

■

Drill 2.3. Let $v(t) = 2t$, $0 \le t < 1$ (a triangular pulse of maximum height 2.) Find the length of this vector for each norm in Eqs. 2.12 through 2.15.

Answer: $\|v\|_\infty = 2,\quad \|v\|_1 = 1,\quad \|v\|_2 = \sqrt{4/3},\quad \|v\|_3 = \sqrt[3]{2}$

Example 2.8. Four norms for the set of all discrete-time waveforms $\{v_i(n)\}$ defined over $0 \le n \le N-1$ are given by

$$\|v\|_\infty = \max_{0 \le n \le N-1} |v(n)| \tag{2.16}$$

$$\|v\|_1 = \sum_{n=0}^{N-1} |v(n)| \tag{2.17}$$

$$\|v\|_2 = \left[\sum_{n=0}^{N-1} |v(n)|^2 \right]^{1/2} \tag{2.18}$$

$$\|v\|_p = \left[\sum_{n=0}^{N-1} |v(n)|^p \right]^{1/p} \tag{2.19}$$

For each type, find the norm of $v_1(n) = 4$ and $v_2(n) = n + 1$, where $0 \le n \le 3$. Use $p = 3$ in the last metric. (Figure 2.4 displays these waveforms.)

Solution:

$$\|v_1\|_\infty = 4 \qquad \|v_2\|_\infty = 4$$

$$\|v_1\|_1 = 4(4) = 16 \qquad \|v_2\|_1 = 4+3+2+1 = 10$$

$$\|v_1\|_2 = \sqrt{4(4^2)} = 8 \qquad \|v_2\|_2 = \sqrt{16+9+4+1} = \sqrt{30}$$

$$\|v_1\|_3 = \sqrt[3]{4(4)^3} = \sqrt[3]{256} \quad \|v_2\|_3 = (64+27+8+1)^{1/3} = (100)^{1/3}$$

■

Example 2.9. Find the length of $y(n)$ in Fig. 2.5 for each norm in Eqs. 2.16 through 2.19. Use $p = 3$ in Eq. 2.19.

Answer: $\|y\|_\infty = 1$, $\|y\|_1 = 4$, $\|y\|_2 = 2$, $\|y\|_3 = \sqrt[3]{4}$.

■

There is a fundamental difference between distance and length. Distance is between two objects. Length applies to only one object. Hence, the two concepts are necessarily distinct. Furthermore, they are not necessarily related. However, they can be related by the metric induced by the norm. Once a norm has been defined for a vector space, a metric is induced by the formula

$$d(v_1, v_2) = \|v_1 - v_2\| \tag{2.20}$$

which sets the distance between two vectors equal to the length (norm) of their difference.

2.4. Inner Product

One more concept in this hierarchy of measures on vector spaces is inner product.

Definition 2.4. Inner Product. The symbol $\langle v_1 | v_2 \rangle$ denotes the operation of extracting a number from a pair of vectors. This is called the *inner* (or *dot*) *product* and must satisfy the following four properties for all scalars $a_i \in C$ and vectors $v_i \in V$.

(i) $\langle v_1 | v_2 \rangle = \langle v_2 | v_1 \rangle^*$.

(ii) $\langle v_1 + v_2 | v_3 \rangle = \langle v_1 | v_3 \rangle + \langle v_2 | v_3 \rangle$.

(iii) $\langle a v_1 | v_2 \rangle = a^* \langle v_1 | v_2 \rangle$.

(iv) $\langle v_1 | v_1 \rangle \geq 0$, and $\langle v_1 | v_1 \rangle = 0$ if and only if $v_1 = v_0$ (the zero vector).

Where a^* denotes the conjugate of the number a.

If the vector space $V = \{v_i\}$ satisfies these properties, it is called an *inner product space*. An inner product space is one in which we can do geometry, because an inner product induces a norm, which in turn induces a metric. The norm is induced by

$$\|v_1\| = \langle v_1 | v_1 \rangle^{1/2} \tag{2.21}$$

which says that in order to determine the norm associated with a given inner product, we find the inner product of a vector with itself and then take the square root. Equations 2.20 and 2.21 establish a sequence that goes from inner product to norm to metric, but not the other way. Incidentally, the "angle" θ between two vectors is given by

$$\cos\theta = \frac{\langle v_1 | v_2 \rangle}{\|v_1\| \, \|v_2\|} \tag{2.22}$$

In the space $V = C^n$ over the field C of complex numbers, the standard inner product of two vectors,

$$v_1 = \begin{bmatrix} a_1 \\ \cdot \\ \cdot \\ \cdot \\ a_n \end{bmatrix} \quad \text{and} \quad v_2 = \begin{bmatrix} b_1 \\ \cdot \\ \cdot \\ \cdot \\ b_n \end{bmatrix}$$

is

$$\langle v_1 | v_2 \rangle = \sum_{i=1}^{n} a_i^* b_i \tag{2.23}$$

which can be written in terms of matrices as $\langle v_1 | v_2 \rangle = v_1^H v_2$, where v_1^H is the Hermitian, the transpose of the conjugate of the matrix v_1:

$$v_1^H = v_1^{*t}$$

For example, if $v_1 = \begin{bmatrix} 1+j & 2 \\ -1 & 3-j2 \\ -3 & 1-j \end{bmatrix}$, then

$$v_1^H = \begin{bmatrix} 1-j & -1 & -3 \\ 2 & 3+j2 & 1+j \end{bmatrix}$$

MATLAB calculates the Hermitian of a matrix with complex entries when you ask for the transpose. This can lead to strange and confusing results when dealing with complex-valued matrices unless you recognize this fact.

The norm induced by this inner product is

$$\|v_1\|_2 = \langle v_1 | v_1 \rangle^{1/2} = \left(\sum_{i=1}^{n} |a_i|^2 \right)^{1/2} \tag{2.24}$$

and the metric (distance between two vectors) is

$$d_2(v_1, v_2) = \|v_1 - v_2\|_2 = \left[\sum_{i=1}^{n} |a_i - b_i|^2 \right]^{1/2} \tag{2.25}$$

These are the usual ways to measure length and distance, and they should be familiar from Euclidean geometry.

Notice that the inner product in Eq. 2.23 multiplies corresponding elements in the two matrices v_1 and v_2. That is, a_1^* multiplies b_1, a_2^* multiplies b_2, and so on. The inner product for $m \times n$ matrices follows this same scheme. For example, let

$$A = \begin{bmatrix} a_{11} & a_{12} \\ a_{21} & a_{22} \end{bmatrix} \qquad B = \begin{bmatrix} b_{11} & b_{12} \\ b_{21} & b_{22} \end{bmatrix}$$

Then

$$\langle A|B \rangle = a_{11}^* b_{11} + a_{12}^* b_{12} + a_{21}^* b_{21} + a_{22}^* b_{22}$$

Another formula for matrix inner product is

$$\langle A|B \rangle = \text{trace}\left(A^H B\right)$$

The *trace* of a matrix is the sum of the elements on the main diagonal. You should convince yourself that these two formulas give the same result.

Example 2.10. Find the inner product, the length, and the distance between the following vectors x and y:

$$x = \begin{bmatrix} 2 & -1 \\ 0 & 1 \\ -2 & -1 \end{bmatrix} \qquad y = \begin{bmatrix} -1 & 0 \\ 2 & 2 \\ 1 & 0 \end{bmatrix}$$

Solution: The sum of the diagonal elements of $x^H y$ is the inner product.

$$x^H y = \begin{bmatrix} 2 & 0 & -2 \\ -1 & 1 & -1 \end{bmatrix} \begin{bmatrix} -1 & 0 \\ 2 & 2 \\ 1 & 0 \end{bmatrix} = \begin{bmatrix} -4 & 0 \\ 2 & 2 \end{bmatrix}$$

The sum of the diagonal elements is $-4 + 2 = -2$, so the inner product is -2. Note that this is the same as summing the products of corresponding elements in the two matrices. Three equivalent formulas for computing the dot product of two matrices of size $M \times N$ are

$$\langle x|y\rangle = \sum_{m=1}^{M}\sum_{n=1}^{N} x_{mn}^{*t} y_{mn} = \text{trace}\left(x^{*t} y\right) = \text{trace}\left(x^{*} y^{t}\right)$$

The inner product induces a norm given by Eq. 2.21. In this case,

$$\|x\| = \langle x|x\rangle^{1/2} = \left(trace(x^H x)\right)^{1/2}$$

and from Eq. 2.20 the distance between x and y is given by

$$d(x,y) = \|x-y\| = \left[\text{trace}\left((x-y)^H (x-y)\right)\right]^{1/2}$$

The MATLAB program given below produced the following results

$$\langle x|y\rangle = -2$$
$$\|x\| = \sqrt{11} = 3.3166$$
$$\|y\| = \sqrt{10} = 3.1623$$
$$d(x,y) = 5$$

```
% Example 2.10
clear

x = [ 2 -1
      0 1
      -2 -1];

y = [ -1 0
      2 2
      1 0];

ip = trace(x'*y)
xnorm = sqrt(trace(x'*x))
ynorm = sqrt(trace(y'*y))
dist = sqrt(trace((x-y)'*(x-y)))
```

■

Example 2.11. Find the inner product, the length, and the distance between the following vectors x and y.

$$x = \begin{bmatrix} 1-j \\ 2 \\ 2+j \end{bmatrix} \qquad y = \begin{bmatrix} 2-j \\ 1+j \\ 1-j2 \end{bmatrix}$$

Solution:

$$\langle x|y\rangle = x^{*t}y = \begin{bmatrix} 1+j & 2 & 2-j \end{bmatrix} \begin{bmatrix} 2-j \\ 1+j \\ 1-2j \end{bmatrix} = 5 - j2$$

$$\|x\| = \left(x^{*t}x\right)^{1/2} = \sqrt{11}$$

$$\|y\| = \left(y^{*t}y\right)^{1/2} = \sqrt{12}$$

$$d(x,y) = \left[(x-y)^{*t}(x-y)\right]^{1/2} = \sqrt{13}$$

Because of their definition, length and distance are always positive real numbers.

■

We deal with four types of signals, and each has its usual definition of inner product.

1. Continuous-time energy signals:

$$\langle v_1 | v_2 \rangle = \int_{-\infty}^{\infty} v_1^*(t) v_2(t)\, dt \tag{2.26}$$

2. Continuous-time power signals:

$$\langle v_1(t) | v_2(t) \rangle = \lim_{T\to\infty} \frac{1}{2T} \int_{-T}^{T} v_1^*(t) v_2(t)\, dt \tag{2.27a}$$

For two periodic signals with identical period T, this reduces to

$$\langle v_1(t) | v_2(t) \rangle = \frac{1}{T} \int_{0}^{T} v_1^*(t) v_2(t)\, dt \tag{2.27b}$$

3. Discrete-time energy signals:

$$\langle v_1(n) | v_2(n) \rangle = \sum_{n=-\infty}^{\infty} v_1^*(n) v_2(n) \tag{2.28}$$

4. Discrete-time power signals:

$$\langle v_1(n) | v_2(n) \rangle = \lim_{N \to \infty} \frac{1}{2N+1} \sum_{n=-N}^{N} v_1^*(n) v_2(n) \tag{2.29a}$$

For two periodic signals with identical period N, this reduces to

$$\langle v_1(n) | v_2(n) \rangle = \frac{1}{N} \sum_{n=0}^{N-1} v_1^*(n) v_2(n) \tag{2.29b}$$

Each of these induces their own norm and metric. For example, for discrete-time energy signals we have

$$\|v(n)\|_2 = \left[\sum_{n=-\infty}^{\infty} |v(n)|^2 \right]^{1/2} \tag{2.30}$$

$$d(v_1, v_2) = \left[\sum_{n=-\infty}^{\infty} |v_1(n) - v_2(n)|^2 \right]^{1/2} \tag{2.31}$$

Notice that the norm squared is the energy in the signal,

$$E = \|v(n)\|_2^2 \tag{2.32}$$

Similarly, for power signals the norm squared is the power in the signal.

Example 2.12. Find the inner product, norm of each signal, and distance between them for the complex exponential signals given by

$$v_1(t) = e^{j\omega t} \qquad v_2(t) = e^{j2\omega t}$$

Solution: These power signals have period $T = 2\pi/\omega$. Therefore,

$$\langle v_1 | v_2 \rangle = \frac{1}{T}\int_0^T e^{-j\omega t} e^{j2\omega t}\, dt = \frac{1}{T}\int_0^T e^{j\omega t}\, dt = 0$$

Note that the integral of a complex exponential over an integer number of periods is zero. The norm of v_1 is given by

$$\|v_1\| = \|e^{j\omega t}\| = \left[\frac{1}{T}\int_0^T e^{-j\omega t} e^{j\omega t}\, dt\right] = 1$$

because $e^{-j\omega t}\, e^{j\omega t} = 1$. Similarly, the norm of v_2 is 1, and the distance between v_1 and v_2 is

$$d(v_1, v_2) = \left[\frac{1}{T}\int_0^T \left(e^{-j\omega t} - e^{-j2\omega t}\right)\left(e^{j\omega t} - e^{j2\omega t}\right) dt\right]^{1/2}$$

$$= \left[\frac{1}{T}\int_0^T (2 - 2\cos\omega t)\, dt\right]^{1/2} = \sqrt{2}$$

■

Drill 2.4. Find the inner product between the signals $x(n)$ and $y(n)$ in Fig. 2.5.

Answer: 0.

2.5. Orthogonality

Two vectors are *orthogonal* if their inner product is 0. It is customary to associate a 90° angle with orthogonality, but the waveforms we deal with have no obvious direction, so you should not look for "right angles" when using orthogonality. For waveforms, orthogonality simply means that $<v_1 | v_2> = 0$. The waveforms in Example 2.12 were orthogonal because their inner product was 0. Orthogonality is a general feature of complex exponential signals of different frequency.

Cauchy-Buniakovsky-Schwartz Inequality

If V is an inner product space, which means we can do geometry, then the following inequality holds for any valid inner product. Let $u, v \in V$. Then

$$|\langle u|v\rangle|^2 \le \langle u|u\rangle\langle v|v\rangle \tag{2.33}$$

with equality if and only if $u = kv$. The proof follows from the properties of the inner product: First, if $v = v_0$, the zero vector, property (iv) of Definition 2.4 gives equality. If $v \neq v_0$, start with

$$0 \le \langle u - \lambda v|u - \lambda v\rangle = \langle u|u\rangle - 2\lambda\langle u|v\rangle + \lambda^2\langle v|v\rangle \tag{2.34}$$

which holds for all λ. Let $\lambda = \langle u|v\rangle / \langle v|v\rangle$. Then Eq. 2.34 gives

$$0 \le \langle u|u\rangle - 2\frac{\langle u|v\rangle^2}{\langle v|v\rangle} + \frac{\langle u|v\rangle^2}{\langle v|v\rangle} = \langle u|u\rangle - \frac{\langle u|v\rangle^2}{\langle v|v\rangle}$$

which gives the CBS inequality in Eq. 2.33.

Example 2.13. Show that the CBS inequality holds for the two matrices in Example 2.10.

$$x = \begin{bmatrix} 2 & -1 \\ 0 & 1 \\ -2 & -1 \end{bmatrix} \qquad y = \begin{bmatrix} -1 & 0 \\ 2 & 2 \\ 1 & 0 \end{bmatrix}$$

Solution: Using the numbers calculated in Example 2.10 gives

$$|\langle x|y\rangle|^2 = 4 \qquad \langle x|x\rangle = 11 \qquad \langle y|y\rangle = 10$$

Therefore, the CBS inequality holds. ■

Puzzle Solution: If you started with a square, you are sunk. Start with a circle. The object is a log with a notch in it, as shown in Fig. 2.6.

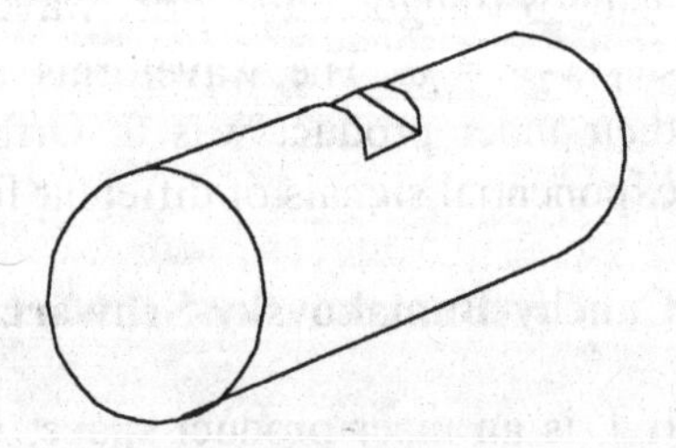

Fig. 2.6. Puzzle Solution.

Chapter 3
Basis and Dimension

The purpose of this chapter is to introduce you to some general notions about vector spaces in preparation for wavelet transforms. These concepts include basis, dimension, orthogonality, and spanning set. These topics build on the idea of a vector space and dot (inner) product from Chapter 2.

Recall that the definition for a vector space is derived from properties of geometric vectors, and that an inner product imposes considerable structure on a vector space. In particular, an inner product $<v_1|v_2>$ induces a norm by the formula

$$\|v\| = \langle v|v\rangle^{1/2}$$

This, in turn, induces a metric or distance measure between two vectors given by

$$d(v_1, v_2) = \|v_1 - v_2\|$$

This makes it possible to do geometry. Once a valid inner product is defined for a vector space, we can measure angles, length, and distance. The angle between v_1 and v_2 is

$$\cos\theta = \frac{\langle v_1|v_2\rangle}{\|v_1\|\|v_2\|}$$

This chapter introduces other aspects of vector spaces, the ideas associated with basis and dimension. You should be warned right away that the dimension of a vector space might not be what you thought it was. For example, the following vector may have any dimension from 1 to N:

$$v = \begin{bmatrix} a_1 \\ a_2 \\ \vdots \\ a_N \end{bmatrix}$$

The concept of dimension applies not to a single vector, but to a set of vectors. All vectors lying in a plane have dimension 2, even though they may have N components.

Another muddled concept is that of basis. A *basis* is a set of vectors with two properties: They are linearly independent, and every vector in the space can be represented as a linear combination of the basis vectors.

Chapter Goals: After completing this chapter, you should be able to do the following:

- Test a set of vectors for independence.
- Find a basis for a given vector space.
- For a given basis, express an arbitrary vector by its coordinates.
- Find the reciprocal basis.

3.1. Linear Independence

To begin at the beginning, define linear independence.

Definition 3.1. A set of vectors $\{v_1, v_2, \ldots, v_n\}$ in the vector space V over the field A is said to be *linearly independent* if the only scalars $\{a_i\}$ such that

$$\sum_{i=1}^{n} a_i v_i = 0 \tag{3.1}$$

are all zero. The set is linearly dependent if it is not independent.

Since only a warped mind can understand this definition on first reading, you may find it better to use the definition for dependence. A set is linearly dependent if any one of the vectors can be expressed as a linear combination of the others. A set is either dependent or independent, with no in-between, so either definition can be used.

Example 3.1. Test the following Euclidean vectors for dependence.

$$v_1 = \begin{bmatrix} 2 \\ 1 \\ -1 \end{bmatrix} \qquad v_2 = \begin{bmatrix} -3 \\ 0 \\ 2 \end{bmatrix} \qquad v_3 = \begin{bmatrix} 2 \\ 4 \\ 0 \end{bmatrix}$$

Solution: If these vectors are dependent, any one of them can be written as a linear combination of the other two. For example,

$$v_1 = av_2 + bv_3$$

or

$$\begin{bmatrix} 2 \\ 1 \\ -1 \end{bmatrix} = a\begin{bmatrix} -3 \\ 0 \\ 2 \end{bmatrix} + b\begin{bmatrix} 2 \\ 4 \\ 0 \end{bmatrix}$$

which gives $a = 0.25$ and $b = -0.5$. Therefore, the vectors are linearly dependent.

■

Drill 3.1. Determine if the following three vectors are independent.

$$v_1 = \begin{bmatrix} 1 \\ 2 \\ 3 \end{bmatrix} \quad v_2 = \begin{bmatrix} -7 \\ 2 \\ -5 \end{bmatrix} \quad v_3 = \begin{bmatrix} -6 \\ 4 \\ 0 \end{bmatrix}$$

Answer: Independent

Example 3.2. Let P_n be the set of all polynomials of degree n or less. Test the following polynomials in P_2 for either dependence or independence.

$$p_1(x) = 1 + 2x$$
$$p_2(x) = 2x + 3x^2$$
$$p_3(x) = 1 + 2x + 3x^2$$

Solution: To test for independence, write

$$a_1p_1(x) + a_2p_2(x) + a_3p_3(x) = 0$$

and see if any nonzero values of a_i make this equation true. If so, the vectors are dependent. If not, they are independent. Plugging into this equation gives

$$a_1(1 + 2x) + a_2(2x + 3x^2) + a_3(1 + 2x + 3x^2) = 0$$

Equating like coefficients gives three equations:

$$\begin{bmatrix} 1 & 0 & 1 \\ 2 & 2 & 2 \\ 0 & 3 & 3 \end{bmatrix} \begin{bmatrix} a_1 \\ a_2 \\ a_3 \end{bmatrix} = \begin{bmatrix} 0 \\ 0 \\ 0 \end{bmatrix} \tag{3.2}$$

The determinant of the coefficient matrix is 6, meaning that the vectors are independent. Since the determinant of the coefficient matrix is not 0, Eq. 3.2 is true only if all the a_i's are 0, which is the condition for independence.

■

In each of the examples above, the coefficient matrix was square. Here is an example to demonstrate the procedure when the coefficient matrix is not square.

Example 3.3. Determine if the following three vectors are independent.

$$v_1 = \begin{bmatrix} -2 \\ -1 \\ 0 \\ 1 \\ 2 \\ 3 \end{bmatrix} \qquad v_2 = \begin{bmatrix} 4 \\ 1 \\ 0 \\ 1 \\ 4 \\ 9 \end{bmatrix} \qquad v_3 = \begin{bmatrix} 2 \\ 1 \\ 0 \\ 1 \\ 2 \\ 3 \end{bmatrix}.$$

Solution: Begin with Eq. 3.1 and write

$$a_1 v_1 + a_2 v_2 + a_3 v_3 = 0 \tag{3.3}$$

Plugging the given vectors into this equation and writing it in matrix form gives the expression in Eq. 3.4. This is an overdetermined set of equations, meaning that there are more equations than unknowns. For this situation, we need to invoke a theorem from linear algebra. (You can probably find this theorem in your calculus book. Look in the chapter on linear algebra.)

$$\begin{bmatrix} -2 & 4 & 2 \\ -1 & 1 & 1 \\ 0 & 0 & 0 \\ 1 & 1 & 1 \\ 2 & 4 & 2 \\ 3 & 9 & 3 \end{bmatrix} \begin{bmatrix} a_1 \\ a_2 \\ a_3 \end{bmatrix} = \begin{bmatrix} 0 \\ 0 \\ 0 \\ 0 \\ 0 \\ 0 \end{bmatrix} \tag{3.4}$$

Theorem. A homogeneous system of m equations in n unknowns (as in Eq. 3.4) has a solution $a_1, a_2, \ldots, a_n$ in which not all the a_i's are 0 if and only if the rank r of the coefficient matrix is less than n.

The rank of a matrix is the largest-dimension square submatrix with nonzero determinant. The first, second, and fourth rows of the coefficient matrix in Eq. 3.4 form a 3 × 3 matrix with nonzero determinant, so the rank of this matrix is 3. The only possible set of a_i's that satisfy the equation are all 0, meaning that the vectors are independent.

■

Drill 3.2. Determine if the following three equations are independent.

$$v_1 = \begin{bmatrix} -2 \\ -1 \\ 0 \\ 1 \\ 2 \\ 3 \end{bmatrix} \quad v_2 = \begin{bmatrix} 4 \\ 1 \\ 0 \\ 1 \\ 4 \\ 9 \end{bmatrix} \quad v_3 = \begin{bmatrix} 0 \\ -1 \\ 0 \\ 3 \\ 8 \\ 15 \end{bmatrix}$$

Answer: Dependent. In fact, $2v_1 + v_2 - v_3 = 0$.

3.2. Basis

A basis is a set of vectors with two properties: Every vector in the space can be represented as a linear combination of the basis vectors, and this representation is unique. The following definition assures that these two conditions are met.

Definition 3.2. Given a vector space V, a subset $\{v_1, v_2, \ldots, v_n\}$ of V forms a *basis* for V if the vectors in the subset are linearly independent and if the addition of any other nonzero vector from V makes the subset dependent.

Let us consider carefully what this means. Figure 3.1a shows vectors in two-dimensional space. Let e_1 and e_2 be candidates for basis vectors. Then v_1 can be expressed as a linear combination of the two, but not v_2. There are not enough basis vectors because e_1 and e_2 are dependent. Figure 3.1b shows three candidate basis vectors e_1, e_2, and e_3. Here there are enough vectors to express v_1, but there is more than one way to do this. We can express v_1 in terms of e_1 and e_2, or e_1 and e_3, or e_2 and e_3. The problem here is that the vectors are again dependent.

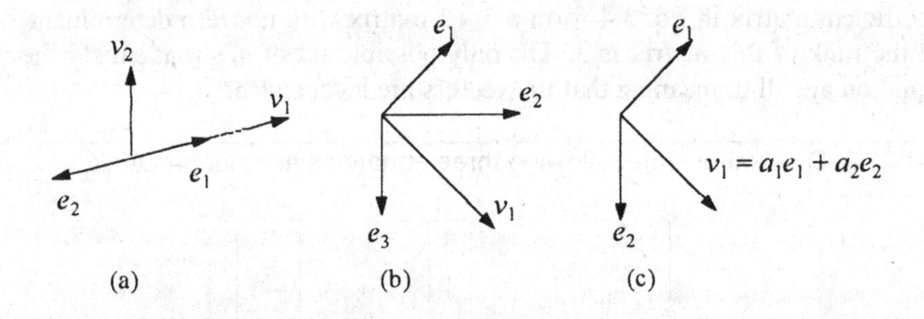

Figure 3.1. Illustrating the idea of a basis.

In order to have a basis there must be enough but not too many vectors. Definition 3.2 specifies just how many this is. Figure 3.1c shows just the right amount for two-dimensional space. Now any vector, such as v_1, can be expressed as a linear combination of e_1 and e_2 in one and only one way.

The vectors in Example 3.1 cannot form a basis for three-dimensional Euclidean space, because they are dependent. The vectors in Example 3.2 can be used for a basis for P_2 because they satisfy Definition 3.2. The vectors in Example 3.3 are independent, but not all vectors in the space can be expressed in terms of these three. There are not enough vectors to form a basis.

Example 3.4. The vectors $\alpha = \{p_1, p_2, p_3\}$ in Example 3.2 form a basis for P_2. This means that we can express any other polynomial from P_2 in only one way in terms of these polynomials. Express

$$p_4(x) = 3 + 2x - 3x^2$$

in terms of this basis.

Solution: Write $p_4(x) = a_1 p_1(x) + a_2 p_2(x) + a_3 p_3(x)$, or

$$\begin{aligned} 3 + 2x - 3x^2 &= a_1(1 + 2x) + a_2(2x + 3x^2) + a_3(1 + 2x + 3x^2) \\ &= a_1 + a_3 + x(2a_1 + 2a_2 + 2a_3) + x^2(3a_2 + 3a_3) \end{aligned}$$

Equating like coefficients gives

$$\begin{aligned} a_1 + a_3 &= 3 \\ 2a_1 + 2a_a + 2a_3 &= 2 \\ 3a_2 + 3a_3 &= -3 \end{aligned}$$

Solving these three equations gives $a_1 = 2$, $a_2 = -2$, $a_3 = 1$. Therefore, $p_4(x)$ can be expressed as

$$p_4(x) = 2p_1(x) - 2p_2(x) + p_3(x)$$

Note that this representation is unique. No other combination of $p_1(x)$, $p_2(x)$, and $p_3(x)$ equals $p_4(x)$.

The components a_1, a_2, and a_3 are called the coordinates of p_4 with respect to the basis α. This is written as

$$[p_4(x)]_\alpha = \begin{bmatrix} 2 \\ -2 \\ 1 \end{bmatrix}$$

We will use this notation in Chapter 4.

■

Drill 3.3. For the basis in Example 3.4, find the coordinate vector for

$$p_5(x) = 5 + 2x - x^2$$

Answer:

$$[p_5(x)]_\alpha = \begin{bmatrix} 4/3 \\ -4 \\ 11/3 \end{bmatrix}$$

3.3 Dimension and Span

The *dimension* of a vector space is the number of vectors in the basis. We often count the number of coordinates and call that the dimension. This may give the correct answer, but the dimension does not depend on a single vector. The dimension depends on all the vectors in the space. Hence, the correct way to determine dimension is to count the number of basis vectors.

And now one more term. Whenever each vector of a vector space V can be expressed as a linear combination of a set of vectors $\{v_1, v_2, \ldots, v_n\}$ of V, we say that the space is spanned by $\{v_1, v_2, \ldots, v_n\}$. This is written

$$V = \text{span}\{v_1, v_2, \ldots, v_n\}$$

Note that this is not the same thing as a basis. There is no requirement that the representation be unique. In other words, a spanning set can contain more vectors than the dimension of the space. Every basis is a spanning set, but so is any other set that contains a basis.

Example 3.5. Find a basis for the set spanned by the vectors

$$v_1 = \begin{bmatrix} 1 \\ 0 \\ 0 \end{bmatrix} \quad v_2 = \begin{bmatrix} 1 \\ -1 \\ 0 \end{bmatrix} \quad v_3 = \begin{bmatrix} 0 \\ 1 \\ -1 \end{bmatrix} \quad v_4 = \begin{bmatrix} 1 \\ 1 \\ 1 \end{bmatrix}$$

Solution: Any three of these vectors will do, say $\{v_1, v_3, v_4\}$. These three form an independent set, and every 3×1 matrix with real entries can be expressed as a linear combination of these vectors. Note that the spanning set $\{v_1, v_2, v_3, v_4\}$ contains more vectors than the basis.

■

Example 3.6. The following set of polynomials spans P_2, the set of all polynomials of degree 2 or less. Find a basis for this set and express the remaining polynomial in terms of this basis.

$$\begin{aligned} x_1(t) &= 1 \\ x_2(t) &= 2t + 2t^2 \\ x_3(t) &= 1 + t + t^2 \\ x_4(t) &= 2t - t^2 \end{aligned}$$

Solution: Note that the first three vectors in this set are dependent. For example,

$$x_3(t) = x_1(t) + \tfrac{1}{2}x_2(t)$$

Therefore, these cannot be used as a basis. Instead, choose $\{x_1, x_2, x_4\}$ as a basis. These three vectors are independent and they span the space, so they are eligible for service as a basis.

In order to express x_3 in terms of this basis, write

$$x_3(t) = a_1 x_1(t) + a_2 x_2(t) + a_3 x_4(t)$$

or

$$1 + t + t^2 = a_1 + a_2\left(2t + t^2\right) + a_3\left(2t - t^2\right)$$

Equating like coefficients gives

$$\begin{aligned} 1 &= a_1 & &\text{or} & a_1 &= 1 \\ t &= 2a_2 t + 2a_3 t & &\text{or} & 2a_2 + 2a_3 &= 1 \\ t^2 &= a_2 t^2 - a_3 t^2 & &\text{or} & a_2 - a_3 &= 1 \end{aligned}$$

In matrix form these equations become

$$\begin{bmatrix} 1 & 0 & 0 \\ 0 & 2 & 2 \\ 0 & 1 & -1 \end{bmatrix} \begin{bmatrix} a_1 \\ a_2 \\ a_3 \end{bmatrix} = \begin{bmatrix} 1 \\ 1 \\ 1 \end{bmatrix}$$

Thus $a_1 = 1$, $a_2 = 3/4$, and $a_3 = -1/4$. (Use MATLAB.)

Check: Does $x_3(t) = a_1 x_1(t) + a_2 x_2(t) + a_3 x_4(t)$?

$$\begin{aligned} x_3(t) &= 1 + \tfrac{3}{4}(2t + t^2) - \tfrac{1}{4}(2t - t^2) \\ &= 1 + t + t^2 \end{aligned}$$

The answer is, yes. Thus, the components of x_3 with respect to the basis $\{x_1, x_2, x_4\}$ are given by

$$[x_3]_{\{x_1, x_2, x_4\}} = \begin{bmatrix} 1 \\ 3/4 \\ -1/4 \end{bmatrix}$$

■

3.4. Reciprocal Bases

Let $\{x_1, x_2, \cdots, x_N\}$ and $\{y_1, y_2, \cdots, y_N\}$ both be sets of the vector space V that satisfy

$$\langle x_i | y_j \rangle = \delta_{ij} \qquad i, j, = 1, 2, \ldots, N \tag{3.5}$$

For example, when $N = 3$, we can write this as

$$\begin{array}{lll} \langle x_1 | y_1 \rangle = 1 & \langle x_1 | y_2 \rangle = 0 & \langle x_1 | y_3 \rangle = 0 \\ \langle x_2 | y_1 \rangle = 0 & \langle x_2 | y_2 \rangle = 1 & \langle x_2 | y_3 \rangle = 0 \\ \langle x_3 | y_1 \rangle = 0 & \langle x_3 | y_2 \rangle = 0 & \langle x_3 | y_3 \rangle = 1 \end{array}$$

Note that $\{x_1, x_2, \ldots, x_N\}$ and $\{y_1, y_2, \ldots, y_N\}$ are two sets of vectors in the same space. They are not in two different spaces. When two such sets have the relationship in Eq. 3.5, they are said to be *reciprocal sets of vectors*.

It is important to note that only one set of vectors can be reciprocal to $\{x_1, x_2, \ldots, x_N\}$. If $\{v_1, v_2, \ldots, v_N\}$ are also reciprocal to $\{x_1, x_2, \ldots, x_N\}$ then

$$\langle x_i | y_j \rangle = \delta_{ij}, \quad i, j, = 1, 2, \cdots, N$$

and

$$\langle x_i | v_j \rangle = \delta_{ij}, \quad i, j, = 1, 2, \cdots, N$$

whence

$$\langle x_i | (y_j - v_j) \rangle = \langle x_i | y_j \rangle - \langle x_i | v_j \rangle$$
$$= 0, \quad i, j, = 1, 2, \cdots, N$$

Therefore, $y_j = v_j$ for all j.

Example 3.7. Let $\alpha = \{\alpha_1, \alpha_2, \alpha_3\}$ be given by

$$\alpha_1 = \begin{bmatrix} 1 \\ 0 \\ 0 \end{bmatrix} \quad \alpha_2 = \begin{bmatrix} 1 \\ 1 \\ 0 \end{bmatrix} \quad \alpha_3 = \begin{bmatrix} 1 \\ 1 \\ 1 \end{bmatrix}$$

These three independent vectors form a basis for R^3. Find the basis set $\beta = \{\beta_1, \beta_2, \beta_3\}$ that is reciprocal to α. Use the usual inner product for $n \times 1$ vectors, $\langle x | y \rangle = x' y$.

Solution: Finding a reciprocal set amounts to applying Eq. 3.5 and cranking through lots of algebra. In the process, we will see that the procedure can be simplified. Start by recognizing that since α is a basis, the vectors $\{\beta_1, \beta_2, \beta_3\}$ can be written in terms of $\{\alpha_1, \alpha_2, \alpha_3\}$.

$$\beta_1 = a_1\alpha_1 + a_2\alpha_2 + a_3\alpha_3 \quad (3.6a)$$
$$\beta_2 = b_1\alpha_1 + b_2\alpha_2 + b_3\alpha_3 \quad (3.6b)$$
$$\beta_3 = c_1\alpha_1 + c_2\alpha_2 + c_3\alpha_3 \quad (3.6c)$$

To solve for the a_i's in the first equation, find the inner product with α_i for $i = 1, 2, 3$.

$$\langle \alpha_1 | \beta_1 \rangle = a_1 \langle \alpha_1 | \alpha_1 \rangle + a_2 \langle \alpha_1 | \alpha_2 \rangle + a_3 \langle \alpha_1 | \alpha_3 \rangle \quad (3.7a)$$

$$\langle \alpha_2 | \beta_1 \rangle = a_1 \langle \alpha_2 | \alpha_1 \rangle + a_2 \langle \alpha_2 | \alpha_2 \rangle + a_3 \langle \alpha_2 | \alpha_3 \rangle \quad (3.7b)$$

$$\langle \alpha_3 | \beta_1 \rangle = a_1 \langle \alpha_3 | \alpha_1 \rangle + a_2 \langle \alpha_3 | \alpha_2 \rangle + a_3 \langle \alpha_3 | \alpha_3 \rangle \quad (3.7c)$$

Numerical values for these inner products are

$$\begin{array}{lll} \langle \alpha_1 | \alpha_1 \rangle = 1 & \langle \alpha_1 | \alpha_2 \rangle = 1 & \langle \alpha_1 | \alpha_3 \rangle = 1 \\ \langle \alpha_2 | \alpha_1 \rangle = 1 & \langle \alpha_2 | \alpha_2 \rangle = 2 & \langle \alpha_2 | \alpha_3 \rangle = 2 \\ \langle \alpha_3 | \alpha_1 \rangle = 1 & \langle \alpha_3 | \alpha_2 \rangle = 2 & \langle \alpha_3 | \alpha_3 \rangle = 3 \end{array} \quad (3.8)$$

Plugging these in and applying Eq. 3.5 gives

$$\begin{bmatrix} 1 & 1 & 1 \\ 1 & 2 & 2 \\ 1 & 2 & 3 \end{bmatrix} \begin{bmatrix} a_1 \\ a_2 \\ a_3 \end{bmatrix} = \begin{bmatrix} \langle \alpha_1 | \beta_1 \rangle \\ \langle \alpha_2 | \beta_1 \rangle \\ \langle \alpha_3 | \beta_1 \rangle \end{bmatrix} = \begin{bmatrix} 1 \\ 0 \\ 0 \end{bmatrix} \quad (3.9)$$

Solving this equation gives

$$\begin{bmatrix} a_1 \\ a_2 \\ a_3 \end{bmatrix} = \begin{bmatrix} 1 & 1 & 1 \\ 1 & 2 & 2 \\ 1 & 2 & 3 \end{bmatrix}^{-1} \begin{bmatrix} 1 \\ 0 \\ 0 \end{bmatrix} = \begin{bmatrix} 2 & -1 & 0 \\ -1 & 2 & -1 \\ 0 & -1 & 1 \end{bmatrix} \begin{bmatrix} 1 \\ 0 \\ 0 \end{bmatrix} = \begin{bmatrix} 2 \\ -1 \\ 0 \end{bmatrix} \quad (3.10)$$

Therefore,

$$\beta_1 = 2\alpha_1 - \alpha_2 = \begin{bmatrix} 1 \\ -1 \\ 0 \end{bmatrix} \quad (3.11)$$

This is the first of the three reciprocal vectors. In order to find β_2 or β_3, repeat the steps above. However, The procedure can be unified by noting that these steps are identical. Consider calculating the b_1, b_2, b_3 coefficients in Eq. 3.6b. The inner products on the right side of Eq. 3.7 remain the same, meaning that the matrix represented by Eqs. 3.8 remains the same. Call this matrix M, so in Eq. 3.9 the matrix M is given by

$$M = \begin{bmatrix} 1 & 1 & 1 \\ 1 & 2 & 2 \\ 1 & 2 & 3 \end{bmatrix} \quad \text{and} \quad M^{-1} = \begin{bmatrix} 2 & -1 & 0 \\ -1 & 2 & -1 \\ 0 & -1 & 1 \end{bmatrix}$$

The only difference is on the left side of Eqs. 3.7. Replace β_1 by β_2. This gives the general form of Eqs. 3.9 and 3.10 as

$$\begin{bmatrix} a_1 \\ a_2 \\ a_3 \end{bmatrix} = M^{-1} \begin{bmatrix} \langle \alpha_1 | \beta_1 \rangle \\ \langle \alpha_2 | \beta_1 \rangle \\ \langle \alpha_3 | \beta_1 \rangle \end{bmatrix} \tag{3.12a}$$

$$\begin{bmatrix} b_1 \\ b_2 \\ b_3 \end{bmatrix} = M^{-1} \begin{bmatrix} \langle \alpha_1 | \beta_2 \rangle \\ \langle \alpha_2 | \beta_2 \rangle \\ \langle \alpha_3 | \beta_2 \rangle \end{bmatrix} \tag{3.12b}$$

$$\begin{bmatrix} c_1 \\ c_2 \\ c_3 \end{bmatrix} = M^{-1} \begin{bmatrix} \langle \alpha_1 | \beta_3 \rangle \\ \langle \alpha_2 | \beta_3 \rangle \\ \langle \alpha_3 | \beta_3 \rangle \end{bmatrix} \tag{3.12c}$$

where

$$M = \begin{bmatrix} \langle \alpha_1 | \alpha_1 \rangle & \langle \alpha_1 | \alpha_2 \rangle & \langle \alpha_1 | \alpha_3 \rangle \\ \langle \alpha_2 | \alpha_1 \rangle & \langle \alpha_2 | \alpha_2 \rangle & \langle \alpha_2 | \alpha_3 \rangle \\ \langle \alpha_3 | \alpha_1 \rangle & \langle \alpha_3 | \alpha_2 \rangle & \langle \alpha_3 | \alpha_3 \rangle \end{bmatrix} \tag{3.13}$$

Plugging numerical values into Eq. 3.12b we get

$$\begin{bmatrix} b_1 \\ b_2 \\ b_3 \end{bmatrix} = \begin{bmatrix} 2 & -1 & 0 \\ -1 & 2 & -1 \\ 0 & -1 & 1 \end{bmatrix} \begin{bmatrix} 0 \\ 1 \\ 0 \end{bmatrix} = \begin{bmatrix} -1 \\ 2 \\ -1 \end{bmatrix}$$

giving

$$\beta_2 = -\alpha_1 + 2\alpha_2 - \alpha_3 = -\begin{bmatrix} 1 \\ 0 \\ 0 \end{bmatrix} + 2\begin{bmatrix} 1 \\ 1 \\ 0 \end{bmatrix} - \begin{bmatrix} 1 \\ 1 \\ 1 \end{bmatrix} = \begin{bmatrix} 0 \\ 1 \\ -1 \end{bmatrix}$$

Similarly,

$$\begin{bmatrix} c_1 \\ c_2 \\ c_3 \end{bmatrix} = \begin{bmatrix} 2 & -1 & 0 \\ -1 & 2 & -1 \\ 0 & -1 & 1 \end{bmatrix}\begin{bmatrix} 0 \\ 0 \\ 1 \end{bmatrix} = \begin{bmatrix} 0 \\ -1 \\ 1 \end{bmatrix}$$

giving

$$\beta_3 = -\alpha_2 + \alpha_3 = -\begin{bmatrix} 1 \\ 1 \\ 0 \end{bmatrix} + \begin{bmatrix} 1 \\ 1 \\ 1 \end{bmatrix} = \begin{bmatrix} 0 \\ 0 \\ 1 \end{bmatrix}$$

Thus, the reciprocal basis is given by

$$\beta = \{\beta_1, \beta_2, \beta_3\} = \left\{ \begin{bmatrix} 1 \\ -1 \\ 0 \end{bmatrix} \begin{bmatrix} 0 \\ 1 \\ -1 \end{bmatrix} \begin{bmatrix} 0 \\ 0 \\ 1 \end{bmatrix} \right\}$$

■

Solving for the reciprocal basis in Eq. 3.6 amounts to solving Eqs. 3.12 and 3.13 for the a_i's, b_i's, and c_i's and then using these in Eq. 3.6.

Note that the $\{\alpha_i\}$ basis is not orthonormal. If it is orthonormal, then M is the identity matrix, thus implying that α and β are identical bases. To see this, combine Eq. 3.12 into one equation:

$$\begin{bmatrix} a_1 & b_1 & c_1 \\ a_2 & b_2 & c_2 \\ a_3 & b_3 & c_3 \end{bmatrix} = M^{-1}\begin{bmatrix} \langle\alpha_1|\beta_1\rangle & \langle\alpha_1|\beta_2\rangle & \langle\alpha_1|\beta_3\rangle \\ \langle\alpha_2|\beta_1\rangle & \langle\alpha_2|\beta_2\rangle & \langle\alpha_2|\beta_3\rangle \\ \langle\alpha_3|\beta_1\rangle & \langle\alpha_3|\beta_2\rangle & \langle\alpha_3|\beta_3\rangle \end{bmatrix} \quad (3.14)$$

The M matrix is the identity matrix because it is orthonormal. Now compare the right matrix to Eq. 3.5 and you can see that it is also the identity matrix, so the matrix on the left is the identity matrix. This means that $a_1 = b_2 = c_3 = 1$ and all other coefficients are zero. Equation 3.6 gives $\beta_1 = \alpha_1$, $\beta_2 = \alpha_2$, and $\beta_3 = \alpha_3$.

Take a closer look at Eq. 3.14. The matrix on the right is the identity matrix by definition, regardless of M. Hence we have the result that the coefficient matrix on the left is M^{-1}. In other words,

$$M^{-1} = \begin{bmatrix} a_1 & b_1 & c_1 \\ a_2 & b_2 & c_2 \\ a_3 & b_3 & c_3 \end{bmatrix} \tag{3.15}$$

This gives one simple formula to find the reciprocal basis β from the original basis α.

$$\beta = \left(M^{-1}\right)^t \alpha \tag{3.16}$$

where M is defined in Eq. 3.13.

Example 3.8. Let $\alpha = \{\alpha_1, \alpha_2\}$ be a basis for R^2 given by

$$\alpha_1 = \begin{bmatrix} 1 \\ 1 \end{bmatrix} \qquad \alpha_2 = \begin{bmatrix} 1 \\ -1 \end{bmatrix}$$

Find the basis $\beta = \{\beta_1, \beta_2\}$ that is reciprocal to α.

Solution: Note that α is orthogonal but not normal, meaning that the length of each basis vector is not 1. Calculating M gives

$$M = \begin{bmatrix} \langle \alpha_1 | \alpha_1 \rangle & \langle \alpha_1 | \alpha_2 \rangle \\ \langle \alpha_2 | \alpha_1 \rangle & \langle \alpha_2 | \alpha_2 \rangle \end{bmatrix} = \begin{bmatrix} 2 & 0 \\ 0 & 2 \end{bmatrix}$$

Hence

$$A = M^{-1} = \begin{bmatrix} 1/2 & 0 \\ 0 & 1/2 \end{bmatrix}$$

Therefore,

$$\beta = \left(M^{-1}\right)^t \alpha = \begin{bmatrix} 1/2 & 0 \\ 0 & 1/2 \end{bmatrix} \begin{bmatrix} \alpha_1 \\ \alpha_2 \end{bmatrix}$$

or

$$\beta_1 = \tfrac{1}{2}\alpha_1 = \begin{bmatrix} 1/2 \\ 1/2 \end{bmatrix}$$

$$\beta_2 = \tfrac{1}{2}\alpha_2 = \begin{bmatrix} 1/2 \\ -1/2 \end{bmatrix}$$

From this example, we may conclude correctly that if the α basis is orthogonal, so is β, and the two basis sets are collinear. If, in addition, α is normal, the two basis sets are identical.

■

Example 3.9. Here is an example using a different vector space. Let P_1 be the set of all polynomials of degree 1 or less. Use as inner product

$$\langle p_1 | p_2 \rangle = \int_0^1 p_1(t) p_2(t)\, dt$$

Find the reciprocal basis for $\alpha = \{\alpha_1, \alpha_2\} = \{1,\ 1+t\}$.

Solution: The matrix M is given by

$$M = \begin{bmatrix} \langle \alpha_1 | \alpha_1 \rangle & \langle \alpha_1 | \alpha_2 \rangle \\ \langle \alpha_2 | \alpha_1 \rangle & \langle \alpha_2 | \alpha_2 \rangle \end{bmatrix} = \begin{bmatrix} 1 & 3/2 \\ 3/2 & 7/3 \end{bmatrix}$$

giving

$$M^{-1} = \begin{bmatrix} 28 & -18 \\ -18 & 12 \end{bmatrix}$$

Therefore,

$$\beta_1 = 28\alpha_1 - 18\alpha_2 = 28 - 18(1+t) = 10 - 18t$$
$$\beta_2 = -18\alpha_1 + 12\alpha_2 = -18 + 12(1+t) = -6 + 12t$$

■

Drill 3.4. Change the inner product to $\langle p_1 | p_2 \rangle = \int_0^2 p_1(t) p_2(t)\, dt$ in Example 3.9 and find the reciprocal basis for $\alpha = \{\alpha_1, \alpha_2\} = \{1,\ 1+t\}$.

Answer: $\beta = \{\beta_1, \beta_2\} = \{3.5 - 3t,\ -1.5 + 1.5t\}$.

Example 3.10. Let V be the set of all discrete-time waveforms of length 4. The set $\{\alpha_i\}$ in Fig. 3.2 is a basis for V. Find the reciprocal basis.

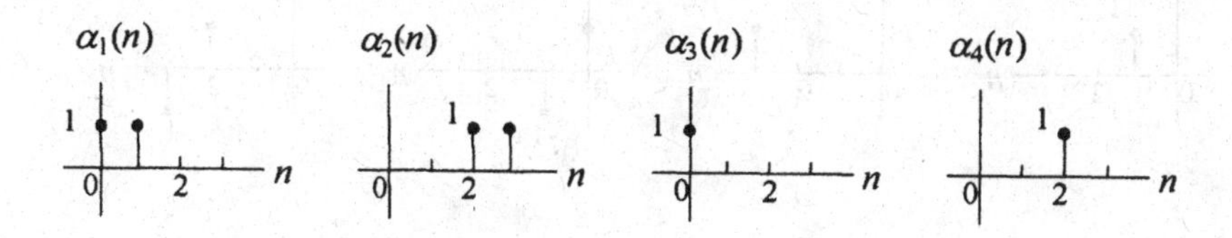

Figure 3.2. The basis α.

Solution: Assuming the usual inner product, given by

$$\langle \alpha_i | \alpha_j \rangle = \sum_{n=0}^{3} \alpha_i(n)\alpha_j(n)$$

the M matrix is given by

$$M = [\langle \alpha_i | \alpha_j \rangle] = \begin{bmatrix} 2 & 0 & 1 & 0 \\ 0 & 2 & 0 & 1 \\ 1 & 0 & 1 & 0 \\ 0 & 1 & 0 & 1 \end{bmatrix}$$

From Eq. 3.13, the reciprocal basis β can be found by

$$\beta = (M^{-1})^t \alpha = \begin{bmatrix} 1 & 0 & -1 & 0 \\ 0 & 1 & 0 & -1 \\ -1 & 0 & 2 & 0 \\ 0 & -1 & 0 & 2 \end{bmatrix} \begin{bmatrix} \alpha_1(n) \\ \alpha_2(n) \\ \alpha_3(n) \\ \alpha_4(n) \end{bmatrix}$$

or

$$\beta_1 = \alpha_1 - \alpha_3 = (1, 1, 0, 0) - (1, 0, 0, 0) = (0, 1, 0, 0)$$
$$\beta_2 = \alpha_2 - \alpha_4 = (0, 0, 1, 1) - (0, 0, 1, 0) = (0, 0, 0, 1)$$
$$\beta_3 = -\alpha_1 + 2\alpha_3 = -(1, 1, 0, 0) + 2(1, 0, 0, 0) = (1, -1, 0, 0)$$
$$\beta_4 = -\alpha_2 + 2\alpha_4 = -(0, 0, 1, 1) + 2(0, 0, 1, 0) = (0, 0, 1, -1)$$

Figure 3.3 shows this reciprocal basis.

■

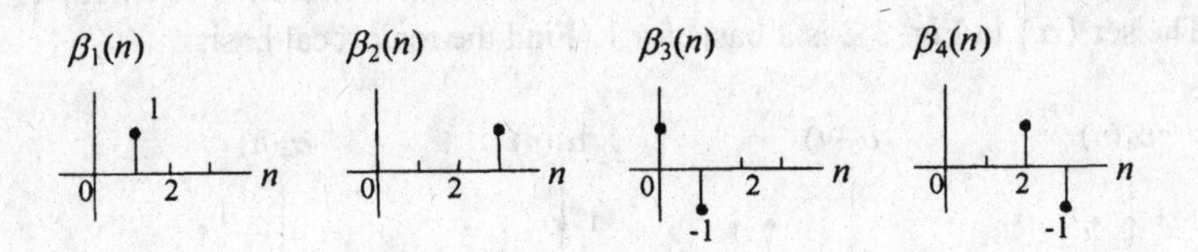

Fig. 3.3. The reciprocal basis β.

Chapter 4
Linear Transformations

Orthonormal bases are commonly used in manipulations of vectors. It simplifies the algebra considerably. This chapter uses more complex bases to illustrate concepts that remain unclear with orthonormal bases. We demonstrate that the coordinates of a vector depend on the basis and give examples for finding these coordinates for vectors in a variety of vector spaces.

The focus is on linear functions (linear transformations) between two finite-dimensional vector spaces. In addition to applying the transformation directly to elements in the domain, a matrix of transformation can operate on the coordinate vectors and produce the same result. In the process, we will discover that a matrix is not simply a rectangular array of numbers, but it is much more. A *matrix* is a representation of a linear transformation between two finite-dimensional vector spaces.

Chapter Goals: After completing this chapter, you should be able to do the following:

- For a given basis, express an arbitrary vector by its coordinates.
- For a given linear transformation and given bases for the domain and codomain, determine the matrix of transformation.

4.1. Component Vectors

Once a basis has been selected for a vector space V, every vector in V may be represented by its components. Let $v \in V$, and let $\beta = \{\beta_1, \beta_2, \ldots, \beta_n\}$ be an ordered basis. (This means that the order is important, i.e., $\{\beta_3, \beta_1, \beta_2, \ldots, \beta_n\}$ is not the same as $\{\beta_1, \beta_2, \ldots, \beta_n\}$.) Then by the definition of basis, v may be represented uniquely as

$$v = a_1\beta_1 + a_2\beta_2 + \cdots + a_n\beta_n \tag{4.1}$$

The $n \times 1$ matrix

$$[v]_\beta = \begin{bmatrix} a_1 \\ a_2 \\ \vdots \\ a_n \end{bmatrix}$$

represents the coordinates of v with respect to β.

Example 4.1. Let P_3 be the set of all polynomials of degree ≤ 3. Then

$$p_1(x) = 5 - x + 2x^2 + 3x^3$$
$$p_2(x) = -1 + x^2$$
$$p_3(x) = 3 + 2x$$

are all in P_3. Let us select $\beta = \{1, x, x^2, x^3\}$ as the basis for P_3. Then the coordinates of these vectors with respect to β are

$$[p_1(x)]_\beta = \begin{bmatrix} 5 \\ -1 \\ 2 \\ 3 \end{bmatrix} \qquad [p_2(x)]_\beta = \begin{bmatrix} -1 \\ 0 \\ 1 \\ 0 \end{bmatrix} \qquad [p_3(x)]_\beta = \begin{bmatrix} 3 \\ 2 \\ 0 \\ 0 \end{bmatrix}$$

Notice that the order for these coordinates is determined by the order of the basis vectors. That is, if the basis is in opposite order, say $\alpha = \{x^3, x^2, x, 1\}$, then the coordinate vectors are given by

$$[p_1(x)]_\alpha = \begin{bmatrix} 3 \\ 2 \\ -1 \\ 5 \end{bmatrix} \qquad [p_2(x)]_\alpha = \begin{bmatrix} 0 \\ 1 \\ 0 \\ -1 \end{bmatrix} \qquad [p_3(x)]_\alpha = \begin{bmatrix} 0 \\ 0 \\ 2 \\ 3 \end{bmatrix}$$

■

Example 4.2. Let $\gamma = \{\gamma_1, \gamma_2, \gamma_3, \gamma_4,\}$ be a basis for P_3 in Example 4.1, where

$$\gamma_1 = 1$$
$$\gamma_2 = 1 + x$$

$$\gamma_3 = 1 + x + x^2$$
$$\gamma_4 = 1 + x + x^2 + x^3$$

Find the coordinates of each vector with respect to γ.

Solution: To find the coordinates of $p_1(x)$, use Eq. 4.1 and write

$$p_1 = a_1\gamma_1 + a_2\gamma_2 + a_3\gamma_3 + a_4\gamma_4$$

or

$$5 - x + 2x^2 + 3x^3 = a_1(1) + a_2(1+x) + a_3(1+x+x^2) + a_4(1+x+x^2+x^3)$$
$$= (a_1 + a_2 + a_3 + a_4) + (a_2 + a_3 + a_4)x + (a_3 + a_4)x^2 + a_4x^3$$

Equating like coefficients produces

$$a_4 = 3 \qquad a_3 = -1 \qquad a_2 = -3 \qquad a_1 = 6 \qquad \text{giving} \quad [p_1(x)]_\gamma = \begin{bmatrix} 6 \\ -3 \\ -1 \\ 3 \end{bmatrix}$$

■

Drill 4.1. Find the coordinate representations for p_2 and p_3 in Example 4.1 with respect to the basis γ.

Answers: $[p_2(x)]_\gamma = \begin{bmatrix} -1 \\ -1 \\ 1 \\ 0 \end{bmatrix} \qquad [p_3(x)]_\gamma = \begin{bmatrix} 1 \\ 2 \\ 0 \\ 0 \end{bmatrix}$

Example 4.3. Let V be the set of all 2×2 matrices over the field of complex numbers. Thus

$$v_1 = \begin{bmatrix} 1+j & 0 \\ -1 & 1-j \end{bmatrix} \qquad v_2 = \begin{bmatrix} 2 & -1 \\ 3 & 0 \end{bmatrix} \qquad v_3 = \begin{bmatrix} -j & 1-j2 \\ j & 0 \end{bmatrix}$$

are all members of this vector space. Choose as bases

$$\beta_1 = \begin{bmatrix} 1 & 0 \\ 0 & 0 \end{bmatrix} \quad \beta_2 = \begin{bmatrix} 0 & 1 \\ 0 & 0 \end{bmatrix} \quad \beta_3 = \begin{bmatrix} 0 & 0 \\ 1 & 0 \end{bmatrix} \quad \beta_4 = \begin{bmatrix} 0 & 0 \\ 0 & 1 \end{bmatrix}$$

Find the coordinates of v_1.

Solution: Using Eq. 4.1, we have

$$\begin{bmatrix} 1+j & 0 \\ -1 & 1-j \end{bmatrix} = a_1 \begin{bmatrix} 1 & 0 \\ 0 & 0 \end{bmatrix} + a_2 \begin{bmatrix} 0 & 1 \\ 0 & 0 \end{bmatrix} + a_3 \begin{bmatrix} 0 & 0 \\ 1 & 0 \end{bmatrix} + a_4 \begin{bmatrix} 0 & 0 \\ 0 & 1 \end{bmatrix}$$

Therefore,

$$\begin{aligned} a_1 &= 1+j \\ a_2 &= 0 \\ a_3 &= -1 \\ a_4 &= 1-j \end{aligned} \qquad \text{or} \qquad [v_1]_\beta = \begin{bmatrix} 1+j \\ 0 \\ -1 \\ 1-j \end{bmatrix}$$

■

Drill 4.2. Find the coordinate matrix for v_2 and v_3 in Example 4.3 with respect to the basis β.

Answers: $$[v_2]_\beta = \begin{bmatrix} 2 \\ -1 \\ 3 \\ 0 \end{bmatrix} \qquad [v_3]_\beta = \begin{bmatrix} -j \\ 1-j2 \\ j \\ 0 \end{bmatrix}$$

Example 4.4. Change basis in Example 4.3. Use γ given by

$$\gamma_1 = \begin{bmatrix} 1 & 0 \\ 0 & 0 \end{bmatrix} \quad \gamma_2 = \begin{bmatrix} 1 & 1 \\ 0 & 0 \end{bmatrix} \quad \gamma_3 = \begin{bmatrix} 1 & 1 \\ 1 & 0 \end{bmatrix} \quad \gamma_4 = \begin{bmatrix} 1 & 1 \\ 1 & 1 \end{bmatrix}$$

Find the coordinates of v_1 with respect to γ.

Solution: Using Eq. 4.1, we have

$$\begin{bmatrix} 1+j & 0 \\ -1 & 1-j \end{bmatrix} = a_1\begin{bmatrix} 1 & 0 \\ 0 & 0 \end{bmatrix} + a_2\begin{bmatrix} 1 & 1 \\ 0 & 0 \end{bmatrix} + a_3\begin{bmatrix} 1 & 1 \\ 1 & 0 \end{bmatrix} + a_4\begin{bmatrix} 1 & 1 \\ 1 & 1 \end{bmatrix}$$

Equating like coefficients gives

$$\begin{aligned} 1+j &= a_1 + a_2 + a_3 + a_4 \\ 0 &= a_2 + a_3 + a_4 \\ -1 &= a_3 + a_4 \\ 1-j &= a_4 \end{aligned}$$

or

$$\begin{aligned} a_4 &= 1-j \\ a_3 &= -2+j \\ a_2 &= 1 \\ a_1 &= 1+j \end{aligned} \qquad \text{giving} \qquad [v_1]_\gamma = \begin{bmatrix} 1+j \\ 1 \\ -2+j \\ 1-j \end{bmatrix}$$

■

Drill 4.3. Find the coordinate matrix for v_2 and v_3 in Example 4.3 with respect to the basis γ.

Answers: $[v_2]_\gamma = \begin{bmatrix} 3 \\ -4 \\ 3 \\ 0 \end{bmatrix}$ $[v_3]_\gamma = \begin{bmatrix} -1+j \\ 1-j3 \\ j \\ 0 \end{bmatrix}$

4.2 Matrices

A matrix is a representation of a linear transformation between two finite-dimensional vector spaces. That bears repeating.

Definition 4.1. A *matrix* is a representation of a linear transformation between two finite-dimensional vector spaces.

What does this mean? For one thing, it means that there is more to matrices than just rectangular arrays of numbers. Let $f: X \to Y$ be a linear transformation, and let X and Y be finite-dimensional vector spaces. Once

the domain and codomain, respectively, a vector $x \in X$ may be represented by its coordinate matrix $[x]_\alpha$. After transformation by f, the resulting vector $y \in Y$ may be represented by its components $[y]_\beta$. This gives two column matrices,

$$[x]_\alpha = \begin{bmatrix} a_1 \\ a_2 \\ \vdots \\ a_n \end{bmatrix} \qquad [y]_\beta = \begin{bmatrix} b_1 \\ b_2 \\ \vdots \\ b_m \end{bmatrix}$$

They are related by the function f. It seems natural that there should exist an $m \times n$ matrix A, called the *matrix of transformation*, such that

$$[y]_\beta = A[x]_\alpha \tag{4.2}$$

This gives us two ways to perform the operation f on $x \in X$.

1. $f: X \to Y: x \mapsto f(x)$
2. $A: [x]_\alpha \to [y]_\beta : [x]_\alpha \mapsto A[x]_\alpha$

The first method operates on x directly to produce y. The second method is indirect. For a given x, find $[x]_\alpha$, multiply this coordinate vector by A to obtain $[y]_\beta$, and then use the basis β to express $[y]_\beta$ as y. Figure 4.1 shows these two parallel methods.

The matrix A depends only on the function f and on the bases α and β. The following algorithm finds the matrix of transformation.

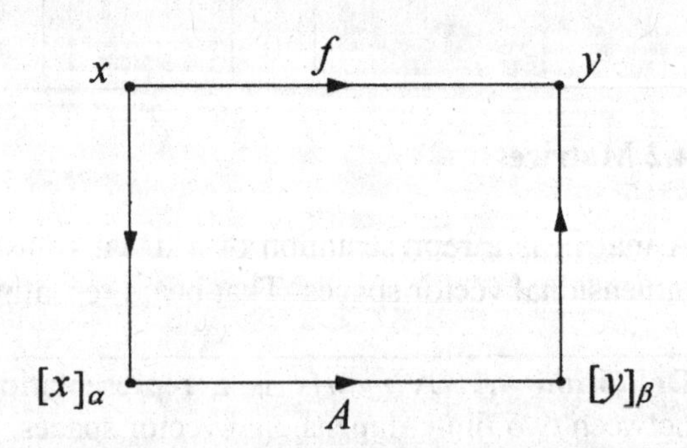

Figure 4.1. Two Parallel Methods.

1. Apply the function f to each basis vector α_i to obtain $y_i = f(\alpha_i)$.
2. Find $[y_i]_\beta$, the coordinates of y_i with respect to the codomain basis β.

3. The matrix of transformation is composed of the column vectors $[y_i]_\beta$. That is,

$$A = \left[[y_1]_\beta \; [y_2]_\beta \; \cdots [y_n]_\beta \right] \tag{4.3}$$

Example 4.5. Let $f: P_2 \to P_3: p(x) \mapsto xp(x)$. That is, f multiplies $p(x) \in P_2$ by x to obtain $y = xp(x)$. For example, if $p(x) = 1 + 2x + x^2$, then $y = xp(x) = x + 2x^2 + x^3$. (You should convince yourself that this transformation is linear.) Choose as bases $\alpha = \{1, x, x^2\}$ and $\beta = \{1, x, x^2, x^3\}$.
(a) Find the matrix of transformation.
(b) Show that f operating on $p(x)$ is equivalent to A operating on the component vector.

Solution: (a) Apply f to each basis vector $\alpha_i \in P_2$.

$$x \times 1 = x$$
$$x \times x = x^2$$
$$x \times x^2 = x^3$$

Express each result by its coordinates in P_3.

$$x = a_1 \times 1 + a_2 \times x + a_3 \times x^2 + a_4 \times x^3 \quad \text{or} \quad [x]_\beta = \begin{bmatrix} 0 \\ 1 \\ 0 \\ 0 \end{bmatrix}$$

$$x^2 = a_1 \times 1 + a_2 \times x + a_3 \times x^2 + a_4 \times x^3 \quad \text{or} \quad [x^2]_\beta = \begin{bmatrix} 0 \\ 0 \\ 1 \\ 0 \end{bmatrix}$$

$$x^3 = a_1 \times 1 + a_2 \times x + a_3 \times x^2 + a_4 \times x^3 \quad \text{or} \quad [x^3]_\beta = \begin{bmatrix} 0 \\ 0 \\ 0 \\ 1 \end{bmatrix}$$

Using these three coordinate vectors as the columns in A gives

$$A = \begin{bmatrix} 0 & 0 & 0 \\ 1 & 0 & 0 \\ 0 & 1 & 0 \\ 0 & 0 & 1 \end{bmatrix}$$

(b) First find the component vector $[p]_\alpha$.

$$p(x) = 1 + 2x + x^2 = a_1 + a_2 x + a_3 x^2$$

Therefore,

$$[p(x)]_\alpha = \begin{bmatrix} a_1 \\ a_2 \\ a_3 \end{bmatrix} = \begin{bmatrix} 1 \\ 2 \\ 1 \end{bmatrix}$$

Then

$$A[p(x)]_\alpha = \begin{bmatrix} 0 & 0 & 0 \\ 1 & 0 & 0 \\ 0 & 1 & 0 \\ 0 & 0 & 1 \end{bmatrix} \begin{bmatrix} 1 \\ 2 \\ 1 \end{bmatrix} = \begin{bmatrix} 0 \\ 1 \\ 2 \\ 1 \end{bmatrix} = [xp(x)]_\beta$$

Therefore,

$$xp(x) = 0(\beta_1) + 1(\beta_2) + 2(\beta_3) + 1(\beta_4) = x + 2x^2 + x^3$$

■

Drill 4.4. Let $p(x) = 2x^2 - x + 3$ in Example 4.5. Therefore,

$$y = xp(x) = 2x^3 - x^2 + 3x$$

Find $[p(x)]_\alpha$, operate on it by A to obtain $[y]_\beta$, and show that this gives y.

Drill 4.5. Change the basis for the domain in Example 4.5 to

$$\alpha = \{1,\ x,\ 1 + x^2\}$$

The basis β remains the same. Find the matrix of transformation A.

Answer:

$$A = \begin{bmatrix} 0 & 0 & 0 \\ 1 & 0 & 1 \\ 0 & 1 & 0 \\ 0 & 0 & 1 \end{bmatrix}$$

Example 4.6. Let $f: P_3 \to P_2: p(x) \mapsto \frac{d}{dx} p(x)$. Choose as bases $\alpha = \{1, x, x^2, x^3\}$ and $\beta = \{1, x, x^2\}$.

(a) Find the matrix of transformation.

(b) Operate on $p(x) = 10 + 3x - 2x^2 + x^3$ both ways to obtain the derivative of $p(x)$.

Solution: (a) Operating on each basis vector in the domain gives

$$\frac{d}{dx} 1 = 0$$
$$\frac{d}{dx} x = 1$$
$$\frac{d}{dx} x^2 = 2x$$
$$\frac{d}{dx} x^3 = 3x^2$$

Express each result by its coordinates in P_2. Obviously, the coordinates of 0 are zero. The other vectors give

$$1 = a_1 \times 1 + a_2 \times x + a_3 \times x^2 \quad \text{or} \quad [1]_\beta = \begin{bmatrix} 1 \\ 0 \\ 0 \end{bmatrix}$$

$$2x = a_1(1) + a_2(x) + a_3(x^2) \quad \text{or} \quad [2x]_\beta = \begin{bmatrix} 0 \\ 2 \\ 0 \end{bmatrix}$$

$$3x^2 = a_1(1) + a_2(x) + a_3(x^2) \quad \text{or} \quad [3x^2]_\beta = \begin{bmatrix} 0 \\ 0 \\ 3 \end{bmatrix}$$

Using these four coordinate vectors as the columns of A gives

$$A = \begin{bmatrix} 0 & 1 & 0 & 0 \\ 0 & 0 & 2 & 0 \\ 0 & 0 & 0 & 3 \end{bmatrix}$$

(b) Find the component vector $[p(x)]_\alpha$. Since

$$p(x) = 10 + 3x - 2x^2 + x^3 = a_1 + a_2 x + a_3 x^2 + a_4 x^3$$

then

$$[p(x)]_\alpha = \begin{bmatrix} a_1 \\ a_2 \\ a_3 \\ a_4 \end{bmatrix} = \begin{bmatrix} 10 \\ 3 \\ -2 \\ 1 \end{bmatrix}$$

Multiplying by A gives

$$A[p(x)]_\alpha = \begin{bmatrix} 0 & 1 & 0 & 0 \\ 0 & 0 & 2 & 0 \\ 0 & 0 & 0 & 3 \end{bmatrix} \begin{bmatrix} 10 \\ 3 \\ -2 \\ 1 \end{bmatrix} = \begin{bmatrix} 3 \\ -4 \\ 3 \end{bmatrix}$$

Therefore,

$$\frac{d}{dt} p(x) = 3(\beta_1) - 4(\beta_2) + 3(\beta_3) = 3 - 4x + 3x^2$$

■

Drill 4.6. Let $p(x) = 2x^2 - x + 3$ in Example 4.6. Therefore,

$$y = \frac{d}{dx} p(x) = 4x - 1$$

Find $[p(x)]_\alpha$, multiply by A to obtain $[y]_\beta$, and show that this gives y.

Example 4.7. Let V be the set of all 2×2 matrices over the field of complex numbers, with $v = \begin{bmatrix} v_{11} & v_{12} \\ v_{21} & v_{22} \end{bmatrix}$. The transformation $f: V \to V$ is given by

$$f(v) = \begin{bmatrix} v_{11} - v_{12} & -v_{12} \\ -v_{21} & v_{22} - v_{21} \end{bmatrix}$$

Choose as bases for both the domain and the codomain

$$\beta_1 = \begin{bmatrix} 1 & 0 \\ 0 & 0 \end{bmatrix} \quad \beta_2 = \begin{bmatrix} 0 & 1 \\ 0 & 0 \end{bmatrix} \quad \beta_3 = \begin{bmatrix} 0 & 0 \\ 1 & 0 \end{bmatrix} \quad \beta_4 = \begin{bmatrix} 0 & 0 \\ 0 & 1 \end{bmatrix}$$

Find the matrix of transformation.

Solution: Apply f to each domain basis vector:

$$f(\beta_1) = \begin{bmatrix} 1 & 0 \\ 0 & 0 \end{bmatrix} \qquad f(\beta_2) = \begin{bmatrix} -1 & -1 \\ 0 & 0 \end{bmatrix}$$

$$f(\beta_3) = \begin{bmatrix} 0 & 0 \\ -1 & -1 \end{bmatrix} \qquad f(\beta_4) = \begin{bmatrix} 0 & 0 \\ 0 & 1 \end{bmatrix}$$

Find the coordinate vectors for each of these with respect to the codomain basis.

$$[f(\beta_1)]_\beta = \begin{bmatrix} 1 \\ 0 \\ 0 \\ 0 \end{bmatrix} \qquad [f(\beta_2)]_\beta = \begin{bmatrix} -1 \\ -1 \\ 0 \\ 0 \end{bmatrix}$$

$$[f(\beta_3)]_\beta = \begin{bmatrix} 0 \\ 0 \\ -1 \\ -1 \end{bmatrix} \qquad [f(\beta_4)]_\beta = \begin{bmatrix} 0 \\ 0 \\ 0 \\ 1 \end{bmatrix}$$

Hence

$$A = \begin{bmatrix} 1 & -1 & 0 & 0 \\ 0 & -1 & 0 & 0 \\ 0 & 0 & -1 & 0 \\ 0 & 0 & -1 & 1 \end{bmatrix}$$

■

Drill 4.7. In Example 4.7, let

$$v_1 = \begin{bmatrix} 3 & 2 \\ -1 & 0 \end{bmatrix}$$

(a) Operate on v_1 by f directly to obtain y.
(b) Find $[f(v_1)]_{\beta}$, multiply by A to obtain $[y]_{\beta}$, and show that this gives y.

Drill 4.8. Let $f: P_2 \to P_3 : p(x) \mapsto \int_0^x p(\lambda)\, d\lambda$. Choose as bases α = $\{1, x, x^2\}$ and $\beta = \{1, x, x^2, x^3\}$. Find the matrix of transformation.

Answer:

$$A = \begin{bmatrix} 0 & 0 & 0 \\ 1 & 0 & 0 \\ 0 & \frac{1}{2} & 0 \\ 0 & 0 & \frac{1}{3} \end{bmatrix}$$

Example 4.8. An LTI system performs a linear operation on the input signal $v(n)$ to produce the output signal $y(n)$. This is one of three requirements that permit the application of a matrix of transformation. The other two are a finite-dimensional domain (set of input signals) and a finite-dimensional codomain (set of output signals). Suppose that a discrete-time LTI system has impulse response $h(n) = \{1, 1, 1\}$ and that the input signals have length 4. This means that the dimension of the input vector space is 4 and the dimension of the output vector space is 6.
(a) Find the matrix of transformation.
(b) Use matrix multiplication to find the response to an input signal $v_1(n) = \{1, 3, -1, 1\}$.
(c) Does the matrix of transformation apply to complex-valued signals? That is, can the response to $v_2(n) = \{1, 2 + j1, 0, 1 - j1\}$ be found by matrix multiplication?

Solution: (a) Let V be the set of all complex-valued input sequences of length 4, and let Y be the set of all output sequences of length 6. Let f stand for the operation performed by the system on the input signal. Then f is described by $f: V \to Y : v(n) \mapsto h(n) * v(n) = y(n)$, where the convolution operation is given by the formula

$$y(n) = \sum_{k=0}^{3} v(k)h(n-k)$$

It would be foolish to choose anything but the usual orthonormal bases for the domain and codomain. Let α and β be the respective bases, where

$$\alpha = \{(1000)\ (0100)\ (0010)\ (0001)\}$$
$$\beta = \{(100000)\ (010000)\ (001000)\ (000100)\ (000010)\ (000001)\}$$

a) To find A, operate on the basis vectors α.

$$h * \alpha_1 = (111000) \Rightarrow [h * \alpha_1]_\beta = [1\ 1\ 1\ 0\ 0\ 0]^t$$
$$h * \alpha_2 = (011100) \Rightarrow [h * \alpha_2]_\beta = [0\ 1\ 1\ 1\ 0\ 0]^t$$
$$h * \alpha_3 = (001110) \Rightarrow [h * \alpha_3]_\beta = [0\ 0\ 1\ 1\ 1\ 0]^t$$
$$h * \alpha_4 = (000111) \Rightarrow [h * \alpha_4]_\beta = [0\ 0\ 0\ 1\ 1\ 1]^t$$

Each vector on the right is a column of A, or

$$A = \begin{bmatrix} 1 & 0 & 0 & 0 \\ 1 & 1 & 0 & 0 \\ 1 & 1 & 1 & 0 \\ 0 & 1 & 1 & 1 \\ 0 & 0 & 1 & 1 \\ 0 & 0 & 0 & 1 \end{bmatrix}$$

You should realize that the basis vectors make it possible to gloss over some of the steps in this process. For example, the output signal $h * \alpha_1 =$ (111000) and the component vector $[1\ 1\ 1\ 0\ 0\ 0]^t$ have the same form, but

the process of translating from the output signal to the component vector would not be so simple for any other basis.
(b) The convolution of $v_1(n) = (1, 3, -1, 1)$ with $h(n) = (1, 1, 1)$ produces $y(n) = (1, 4, 3, 3, 0, 1)$. To produce this result by matrix multiplication, first find the component vector for v_1 with respect to the basis α. This gives the 4×1 matrix $[v_1]_\alpha = [1 \;\; 3 \;\; -1 \;\; 1]^t$. Multiply by A to obtain

$$A[v_1]_\alpha = \begin{bmatrix} 1 & 0 & 0 & 0 \\ 1 & 1 & 0 & 0 \\ 1 & 1 & 1 & 0 \\ 0 & 1 & 1 & 1 \\ 0 & 0 & 1 & 1 \\ 0 & 0 & 0 & 1 \end{bmatrix} \begin{bmatrix} 1 \\ 3 \\ -1 \\ 1 \end{bmatrix} = \begin{bmatrix} 1 \\ 4 \\ 3 \\ 3 \\ 0 \\ 1 \end{bmatrix}$$

This is the component vector for the output sequence $y(n)$.
c) Of course, it works just fine on complex-valued signals.

$$A[v_2]_\alpha = \begin{bmatrix} 1 & 0 & 0 & 0 \\ 1 & 1 & 0 & 0 \\ 1 & 1 & 1 & 0 \\ 0 & 1 & 1 & 1 \\ 0 & 0 & 1 & 1 \\ 0 & 0 & 0 & 1 \end{bmatrix} \begin{bmatrix} 1 \\ 2+j1 \\ 0 \\ 1-j1 \end{bmatrix} = \begin{bmatrix} 1 \\ 2+j2 \\ 2+j2 \\ 3 \\ 1-j1 \\ 1-j1 \end{bmatrix}$$

The output signal is $y(n) = \{1 \;\; 2+j2 \;\; 2+j2 \;\; 3 \;\; 1-j1 \;\; 1-j1\}$.

■

Drill 4.9. Suppose that the system impulse response in Example 4.8 is $h(n) = \{1, 1, -1\}$. Find the matrix of transformation.

Answer:
$$A = \begin{bmatrix} 1 & 0 & 0 & 0 \\ 1 & 1 & 0 & 0 \\ -1 & 1 & 1 & 0 \\ 0 & -1 & 1 & 1 \\ 0 & 0 & -1 & 1 \\ 0 & 0 & 0 & -1 \end{bmatrix}$$

Drill 4.10. Let *Pn* be the set of all polynomials of degree *n* or less, and let $\alpha = \{1, x, x^2, ..., x^n\}$ be the basis for P_n. Suppose the linear transformation $f: P_2 \to P_3$ has the matrix of transformation given by

$$A = \begin{bmatrix} 1 & 0 & 0 \\ 1 & 1 & 0 \\ 0 & 1 & 1 \\ 0 & 0 & 1 \end{bmatrix}$$

Find the action of *f* on $p(x) = 3 - x + 2x^2$.

Answer: $f[p(x)] = 3 + 2x + x^2 + 2x^3$

Chapter 5
Sampling Theorem

Of all the theorems connected with signal processing, one of the most remarkable is the *sampling theorem.* It relates continuous-time signals to discrete-time signals. It states that if samples are placed close enough together, the continuous-time signal can be recovered *exactly* from its samples. This word *exactly* is what makes the theorem remarkable. Intuitively, one might think that a continuous-time function could be approximated with less and less error as its samples become closer together, thus reducing the error to zero as the number of samples approaches infinity. However, this is not the case. The sampling theorem establishes a threshold called the *minimum sampling rate*. At any rate faster than this threshold there is enough information to recover the continuous-time signal with no error.

Of course, there is no such thing as a free meal. The continuous-time signal must be band-limited, meaning that it can only wiggle so much. If it wiggles any faster, then its bandwidth increases, and the sampling rate must be increased. The signal bandwidth establishes the minimum sampling rate.

Chapter Goals: After completing this chapter, you should be able to do the following:

- Determine the Nyquist sampling rate for a given signal bandwidth.
- Determine the bandwidth of an antialiasing filter for a given sampling rate.
- On paper, convert a sampled signal into a PCM signal.
- Calculate the quantization noise power for a uniformly quantized signal.

5.1. Nyquist Rate

Sampling is the mathematical operation of multiplying a signal $v(t)$ by an impulse train, as depicted in Fig. 5.1. The product $v(t)p(t)$ produces the sampled signal $v_s(t)$.

$$v_s(t) = v(t)\,p(t) \tag{5.1}$$

This multiplication in the time domain corresponds to convolution in the frequency domain, or

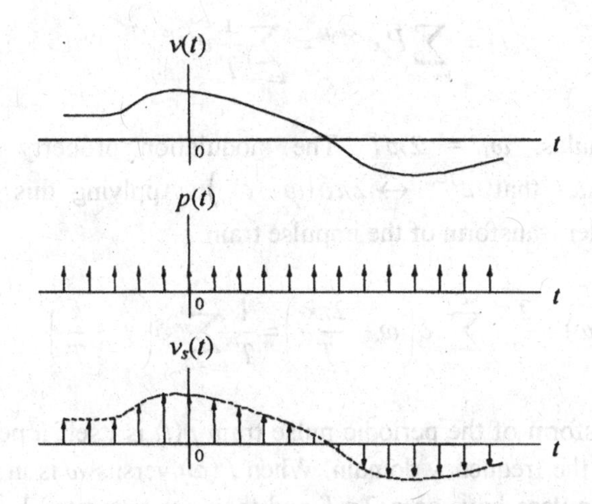

Figure 5.1. The functions $v(t)$ and $p(t)$, and the sampled signal $v_s(t)$.

$$V_s(\omega) = V(\omega) * P(\omega) \tag{5.2}$$

In order to find the spectrum for the impulse train $P(\omega)$, write $p(t)$ as the sum of impulses.

$$p(t) = \sum_{n=-\infty}^{\infty} \delta(t - nT)$$

Note that this pulse train has period T.

Of the several methods for finding the Fourier transform of $p(t)$, the one that probably makes the most sense is first to represent $p(t)$ in a different form. Express $p(t)$ by its Fourier series. Recall that the series coefficients for a continuous-time power signal $p(t)$ are given by

$$P_k = \frac{1}{T}\int_{t'}^{t'+T} p(t) e^{-jk\omega_1 t}\, dt$$

where t' is a value of time (initial time) and T is the period. With $t' = -T/2$ this gives $P_k = 1/T$ since $p(t) = \delta(t)$ in the interval of integration. Thus, the Fourier series of $p(t)$ is

$$p(t) = \sum_{k=-\infty}^{\infty} P_k e^{jk\omega_1 t} = \sum_{k=-\infty}^{\infty} \frac{1}{T} e^{j2\pi kt/T} \tag{5.3}$$

In both formulas, $\omega_1 = 2\pi/T$. The modulation property of Fourier transforms states that $e^{j\omega_0 t} \leftrightarrow 2\pi\delta(\omega - \omega_0)$. Applying this to Eq. 5.3 gives the Fourier transform of the impulse train.

$$P(\omega) = \frac{2\pi}{T} \sum_{k=-\infty}^{\infty} \delta\left(\omega - \frac{2\pi k}{T}\right) = \frac{1}{T} \sum_{k=-\infty}^{\infty} \delta\left(f - \frac{k}{T}\right) \tag{5.4}$$

Thus, the transform of the periodic pulse train $p(t)$ is itself a periodic train of impulses in the frequency domain. When $P(\omega)$ versus ω is in radians per second, the impulses have area $2\pi/T$ and they are separated by $\omega_s = 2\pi/T$ radians per second. When $P(f)$ versus the frequency f is in hertz, the impulses have area $1/T$ and they are separated by $f_s = 1/T$ hertz.

Since $v_s(t)$ is the product $v(t)p(t)$, the transform $V_s(\omega)$ is the convolution of $V(\omega)$ with $P(\omega)$. Figure 5.2 shows the two signals to be convolved in the frequency domain. Assume an arbitrary shape for $V(\omega)$ and convolve this with $P(\omega)$ to produce the spectrum $V_s(\omega)$. Let us assume that the highest frequency in $V(\omega)$ is ω_m. Then Fig. 5.2c reveals that if $\omega_s > 2\omega_m$ there will be no overlap between adjacent components in $V_s(\omega)$. This is the condition necessary for recovery of the original signal $v(t)$ from its samples by passing $v_s(t)$ through a low-pass filter (the dotted line in Fig. 5.2c.) The minimum rate $2\omega_m$ is called the *Nyquist rate*.

Figure 5.3 illustrates the situation when the samples are too far apart, meaning that $\omega_s < 2\omega_m$. The frequency components overlap, causing "interference." This is called *aliasing*, and it results from periodic sampling at a rate below the Nyquist rate.

Notice that after sampling, the signal $v(n)$ is a discrete-time signal. After all, the purpose of sampling is so that we can represent the signal by a sequence of numbers. Thus, the spectrum is a function of discrete frequency, and it is periodic with a period of 2π. This means that we can replace the plot of $V_s(\omega)$ in Fig. 5.2 by a plot of $V_s(\Omega)$. In that case, the value of ω_s in Fig. 5.2 corresponds to 2π. Let us see if this agrees with previous experience.

First, the units on the frequency variables and the sampling rate are given by

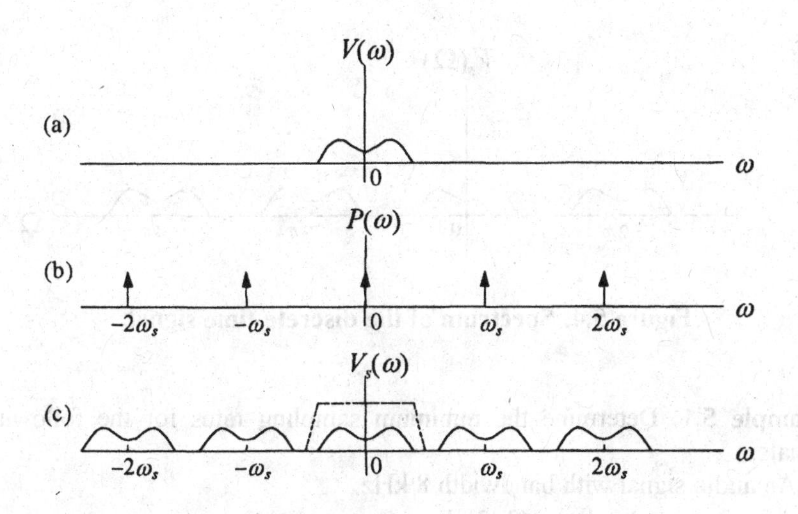

Figure 5.2. Convolving $V(\omega)$ and $P(\omega)$.

ω (radians/second)
Ω (radians/sample)
r_s (samples/second)

Therefore, to make the units agree, the relation must be

$$\omega(\text{radians/second}) = \Omega(\text{radians/sample}) \times r_s(\text{samples/second})$$

Since $\omega_s = 2\pi/T = 2\pi r_s$, then indeed $\Omega_s = \omega_s/r_s = 2\pi$. Therefore, the spectrum of the discrete-time signal looks like that in Fig. 5.4.

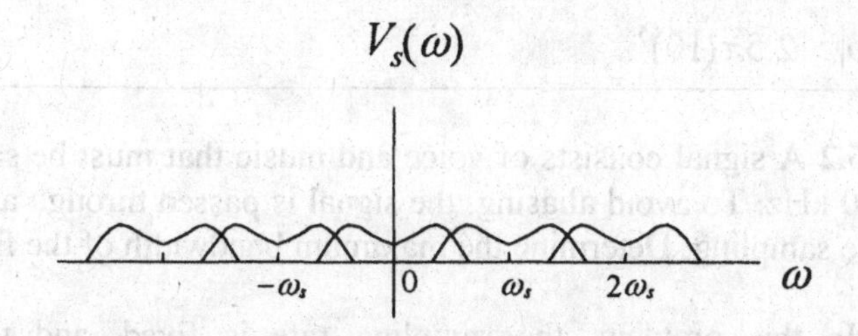

Figure 5.3. The spectrum when $\omega_s < 2\omega_m$.

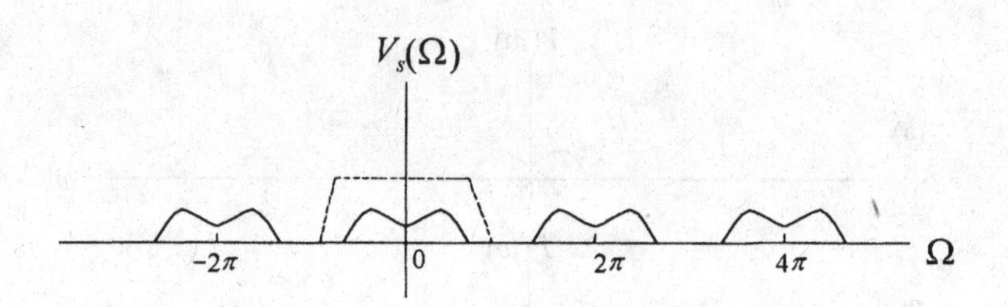

Figure 5.4. Spectrum of the discrete-time signal.

Example 5.1. Determine the minimum sampling rates for the following signals:
(a) An audio signal with bandwidth 8 kHz.
(b) A sinusoid $v(t) = 10\sin(2\pi 20t)$.
(c) A mixture of signal and noise, where the signal is bandlimited to 10 kHz, and the noise is white (i.e., has infinite bandwidth).

Solution: (a) $2f_m = 16$ kHz.
(b) The sinusoid has frequency 20 Hz, so the sampling rate should be at least 40 samples/second.
(c) The minimum sampling rate for this signal is 20 kHz. Since samples consist of signal plus noise for any sampling rate, the noise has no influence on the minimum sampling rate. Therefore, the answer is $f_s \geq 20$ kHz.

■

Drill 5.1. A signal $w(t)$ is known to be determined uniquely by its samples when the sampling frequency is $\omega_s = 5\pi(10)^5$. For what values of ω is $W(\omega)$ equal to zero?

Answer: $|\omega| > 2.5\pi(10)^5$.

Example 5.2 A signal consists of voice and music that must be sampled at a rate of 10 kHz. To avoid aliasing, the signal is passed through a low-pass filter before sampling. Determine the maximum bandwidth of the filter.

Solution: In this problem, the sampling rate is fixed, and the signal bandwidth must be restricted to avoid aliasing. To accomplish this, pass the signal through a low-pass filter with a bandwidth that is small enough to avoid aliasing at the given rate of 10 kHz.

Figure 5.5a shows the spectrum of a signal with bandwidth 5 kHz that is sampled at the Nyquist rate of 10 kHz. An ideal filter with bandwidth f_c = 5 kHz would allow this situation, but practical filters cannot have the sharp transition band required here. There must be some separation between spectral components, as shown in Fig. 5.5b, to allow for the transition band of a practical filter. Practical sampling rates should be about 1.2 times the Nyquist rate to allow for the filter transition band. This means that

$$f_s = 2.4 f_m$$

Since f_s is fixed in this problem, then

$$f_m = \frac{f_s}{2.4} = \frac{10^4}{2.4} = 4.167 \text{ kHz}$$

Thus, the filter should have a bandwidth of 4.167 kHz. ■

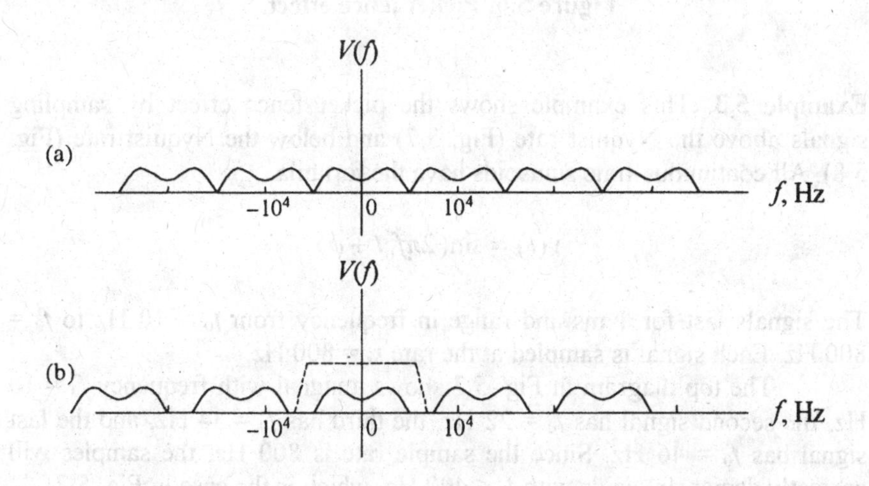

Figure 5.5. Nyquist and practical sampling rate.

Drill 5.2. Suppose that a signal consists of the sum of two sinusoids. Let $v(t) = \cos(6\pi t) + \sin(10\pi t)$.
(a) What is the Nyquist sampling rate?
(b) What sampling rate should be used in practical applications?

Answer: (a) 10 Hz. (b) 12 Hz.

Figure 5.6 shows the "picket fence" effect. Knowing only the samples is like looking through a picket fence. We cannot determine whether the given sample values came from sampling the low-frequency signal or the high-frequency signal. This figure illustrates aliasing clearly. Whenever a high-frequency signal is sampled below its sampling rate, there is another low-frequency signal that fits the samples. The following example illustrates this same effect in a different way.

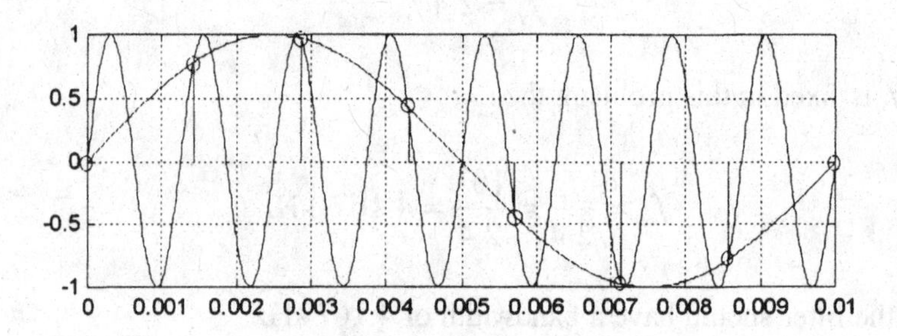

Figure 5.6. Picket fence effect.

Example 5.3. This example shows the picket fence effect by sampling signals above the Nyquist rate (Fig. 5.7) and below the Nyquist rate (Fig. 5.8). All continuous-time sinusoids have the formula

$$v(t) = \sin(2\pi f_0 t + \phi)$$

The signals last for 1 ms and range in frequency from f_0 = 10 Hz to f_0 = 800 Hz. Each signal is sampled at the rate r_s = 800 Hz.

The top diagram in Fig. 5.7 shows a signal with frequency f_0 = 10 Hz, the second signal has f_0 = 22 Hz, the third has f_0 = 34 Hz, and the last signal has f_0 = 46 Hz. Since the sample rate is 800 Hz, the samples will correctly depict sinusoids with $f_0 < 400$ Hz, which is the case in Fig. 5.7.

Contrast this with the four sinusoids in Fig. 5.8. As before, each signal lasts for 1 ms and is sampled at the rate r_s = 800 Hz. However, the sinusoidal frequencies range from 732 to 800 Hz. Notice that the waveforms appear similar to those in Fig. 8.1.7, although the sinusoids in Fig. 5.8 have much higher frequencies. This same effect makes spoke wheels appear to turn backwards in the movies.

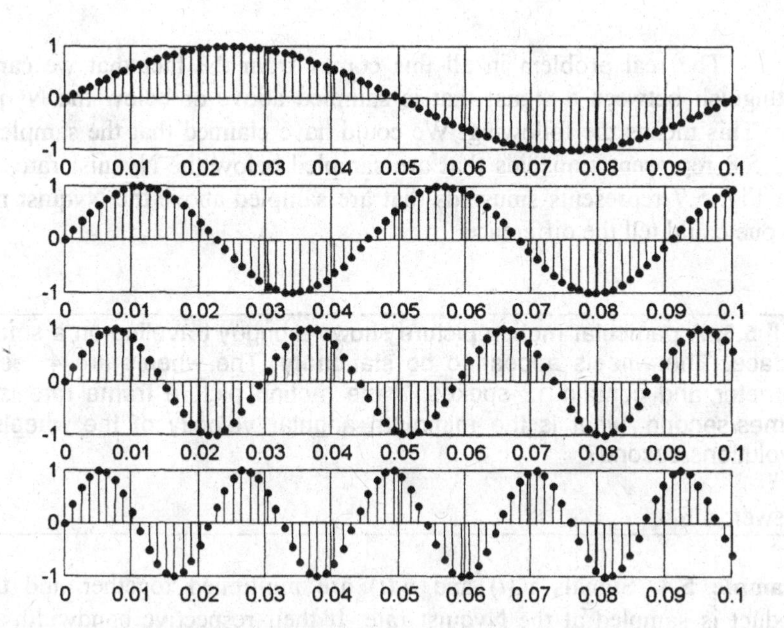

Figure 5.7. Sampling above the Nyquist rate.

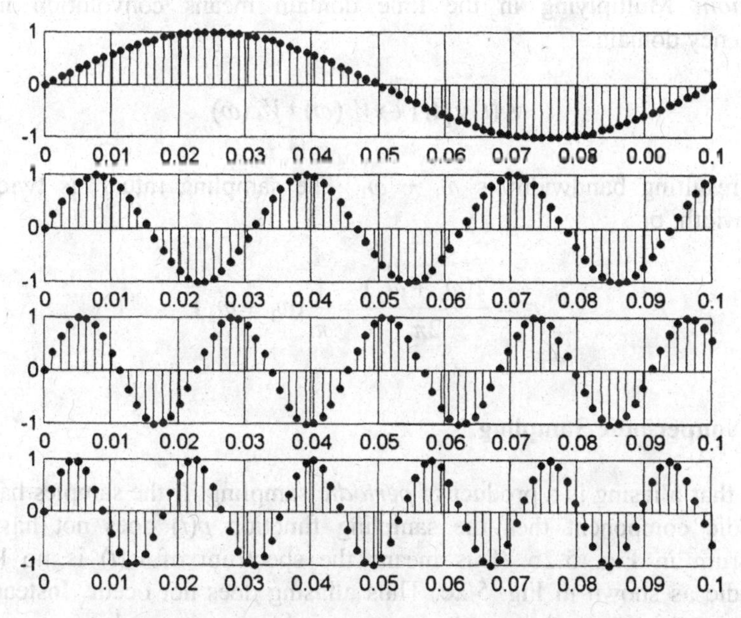

Figure 5.8. Sampling below the Nyquist rate.

The real problem in all this comes from the fact that we cannot distinguish between a signal that is sampled above or below the Nyquist rate. This means the following: We could have claimed that the samples in Fig. 5.8 represent sinusoids that are sampled below the Nyquist rate, and that Fig. 5.7 represents sinusoids that are sampled above the Nyquist rate. No one could tell the difference.

■

Drill 5.3. A particular motion picture shows a buggy traveling on a smooth surface. The wheels appear to be stationary. The wheels are 4 feet in diameter and contain 12 spokes. If the motion picture frame rate is 16 frames/second, what is the minimum angular velocity of the wheels in revolutions/second?

Answer: 4/5.

Example 5.4. Signals $v_1(t)$ and $v_2(t)$ are multiplied together and their product is sampled at the Nyquist rate. If their respective bandwidths are ω_1 and ω_2, what is the sampling rate?

Solution: Multiplying in the time domain means convolution in the frequency domain.

$$v_1(t)v_2(t) \leftrightarrow V_1(\omega) * V_2(\omega)$$

The resulting bandwidth is $\omega_1 + \omega_2$. The sampling rate r_s is twice the bandwidth, or

$$r_s = \frac{2(\omega_1 + \omega_2)}{2\pi} = \frac{1}{\pi}(\omega_1 + \omega_2)$$

■

5.2. Nonperiodic Sampling

Note that aliasing is a product of *periodic* sampling. If the samples have no periodic component then the sampling function $p(t)$ does not have the spectrum in Fig. 5.2b. This means the spectrum of $v_s(t)$ is no longer periodic as shown in Fig. 5.2c. Thus aliasing does not occur. Instead, the spectrum of $v(t)$ is distorted by a nonperiodic sampling scheme, making it difficult, if not impossible, to recover $v(t)$ from its samples. This does not mean that sampling should be periodic in all applications. If the original signal is to be recovered, the sampling should be periodic. Otherwise, you are free to choose between periodic sampling and some other method. This is illustrated in the following Example.

Example 5.5. Here are the results of an experiment performed to demonstrate periodic versus nonperiodic sampling. Suppose that a low-frequency signal of interest is imbedded in an environment of high-frequency signals. All signals may be detected from their samples by sampling at a fast rate, but if uniform sampling is done at an intermediate rate, the high frequencies will be mistaken for low frequencies due to aliasing. Figure 5.9 shows portions of two sinusoids at frequencies $\omega_1 = 2$ and $\omega_2 = 27.45\omega_1$, along with their sum.

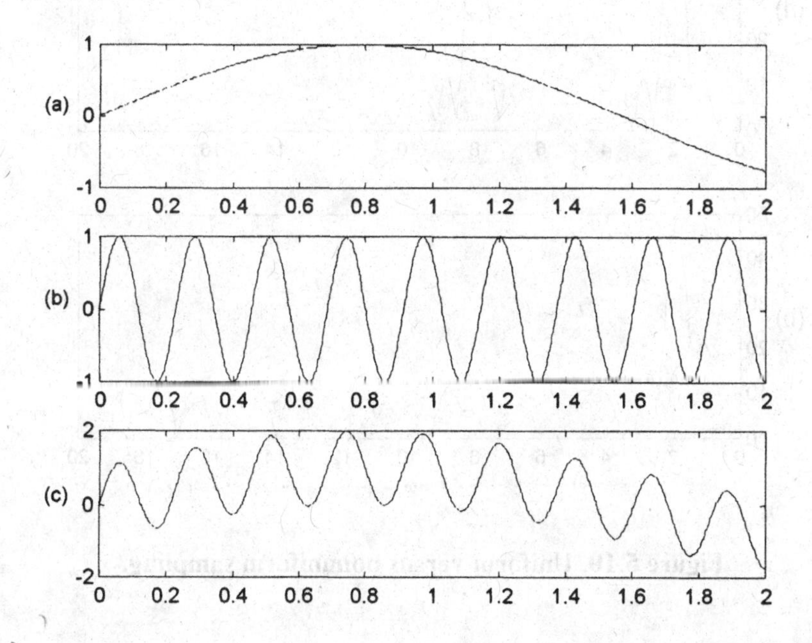

Figure 5.9. The signals in the nonuniform sampling experiment.

Two experiments on the sum, one using uniform sampling and the second using samples whose spacing increases with time, demonstrate the difference. First, sample the sum of these two sinusoids at a uniform rate of 10 Hz, or $31.42f_1$. This intermediate rate is between $2f_1$ and $2f_2$. Figure 5.10a shows the magnitude spectrum resulting from this scheme. The frequency component at $5.97\omega_1$ part(a) is the aliased frequency due to the difference $31.42\omega_1 - 27.45\omega_1$. This component is absent from the frequency plot in Fig. 5.10b, representing non-uniform sampling. Note that there is considerable noise at all frequencies in part(b), once again demonstrating that there is no such thing as a free meal. Although aliasing is eliminated, nonuniform sampling results in more noise. ■

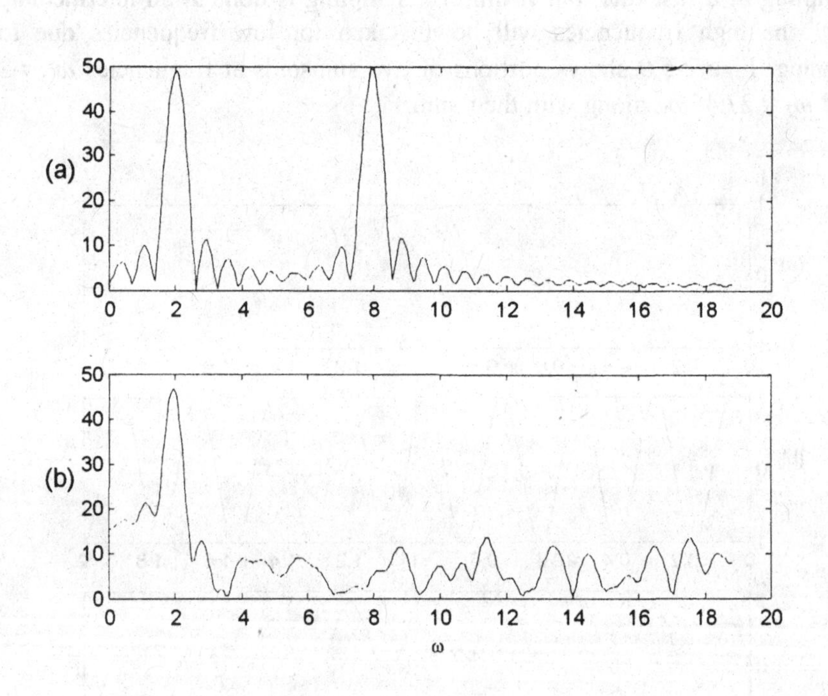

Figure 5.10. Uniform versus nonuniform sampling.

5.3. Quantization and Pulse Code Modulation

Sampling is ubiquitous in our modern life. In making a long-distance telephone call, chances are the voice signals are sampled, quantized, encoded, multiplexed, and transmitted along with many other signals over the same channel. At the receiving end these signals are separated, decoded, and converted back into an analog signal in such a manner that telephone users never realize that their voice signals endured such processing. This processing includes some form of code, usually pulse code modulation (PCM). The following presents the ideas behind quantization and PCM.

The digital computer stores numbers in binary form. There are 2^B different combinations of B binary digits, which means that any sample of an analog signal must be quantized into 2^B levels if we use B digits for each number. For example, if $B = 3$, there are $2^3 = 8$ quantization levels. Figure 5.11 shows 13 samples of a signal with eight quantization levels. Each quantized value is represented with only three binary digits, 000 through 111. This process of converting quantization levels into binary digits is called *pulse code modulation*.

Drill 5.4. Suppose that an audio signal is band limited to 20 kHz. The signal is sampled, quantized, and binary coded to obtain a PCM signal.
(a) Determine the sampling rate if the signal is to be sampled at a rate 20% above the Nyquist rate.
(b) If the samples are transmitted using 16 bits/sample, how many quantization levels are used?
(c) Determine the bit rate (binary digits per second) required to encode the signal.

Answer: (a) 48,000 samples/sec. (b) 65,536 levels. (c) 768,000 bits/sec.

Quantization induces error, labeled *quantization noise*. For example, suppose that the waveform in Fig. 5.11 varies over a 10-volt range. Then each three-digit binary number represents any voltage in a 1.25-volt range. (There are only eight levels to cover 10 volts, so 10/8 = 1.25.) If the system quantizes to the center of each level, the error can range from −0.625 to 0.625.

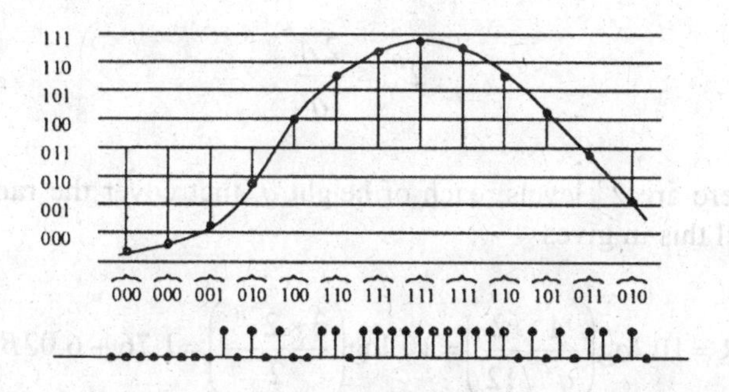

Figure 5.11. Pulse code modulation.

Let q represent the range for each quantization level. (In this example, $q = 1.25$.) Assume that the quantization error for each sample, e, is random with uniform distribution. This means that one value of error between the limits $-q/2 \le e \le q/2$ is just as likely as any other value in this range. In this case, the quantization noise power is given by

$$P_q = \int_{-q/2}^{q/2} e^2 \left(\frac{1}{q}\right) de = \frac{q^2}{12} \tag{5.5}$$

Where did this formula originate? This is the mean square value of a random variable with uniform distribution over the range $(-q/2, q/2)$. The error is distributed uniformly over this interval, so this is the formula for the quantization noise power.

Example 5.6. Suppose that a sinusoid of amplitude A is quantized and sampled with quantization range q. The signal is given by

$$v(t) = A \sin \omega t$$

Find the signal to quantization noise power in dB, or

$$\text{SNR} = 10 \log \frac{P_s}{P_q}$$

Solution: The signal power is $P_s = A^2/2$. The quantization noise power is $P_q = q^2/12$. The signal amplitude A and the quantization range q are related by

$$2^B = \frac{2A}{q}$$

That is, there are 2^B levels, each of height q, that cover the range $(-A, A)$. Plugging all this in gives

$$\text{SNR} = 10 \log\left(\frac{A^2/2}{q^2/12}\right) = 10 \log\left(\frac{3 \times 2^{2B}}{2}\right) = 1.76 + 6.02B \text{ dB}$$

■

5.4. Companding

Figure 5.11 illustrates uniform quantization. If the waveform varies over a 10-volt range, and there are eight quantization levels, then each quantization level represents any voltage in a 1.25-volt range. A small change in this scheme can improve the signal-to-noise ratio for certain signals. Suppose that the signal to be quantized spends most of the time in a small voltage range, say from −2 to +2 volts out of the possible range of −5 to +5 volts. Then it would make sense to have more quantization levels in the range −2 to +2 volts and fewer quantization levels in the ranges from −5 to −2 volts and from +2 to +5 volts. Figure 5.12 shows the altered quantization scheme. Note that the quantization levels are closer together near 0 volts, and they are farther apart near +5 and −5 volts. Therefore, the quantization error will be smaller where the signal level is near zero, and larger where the signal level is large (+5 or −5 volts).

This nonuniform quantization tends to make the signal-to-noise (S/N) ratio equal for all input levels. That is, if the speaker has a soft voice, more quantization levels at low amplitude increase the S/N ratio. If the speaker is loud, this scheme decreases the S/N ratio. From Eq. 5.5 the quantization noise power is $q^2/12$, where q is the quantization step size. If q changes with magnitude, the noise due to quantization changes with magnitude. This scheme can either increase or decrease the S/N ratio, and smaller q at low signal levels increases S/N for low signals.

Nonuniform quantization changes the quantizer but leaves the signal alone. An equivalent scheme changes the signal and uses a uniform quantizer. This is called *companding* because the signal is both compressed and expanded in the transmitter.

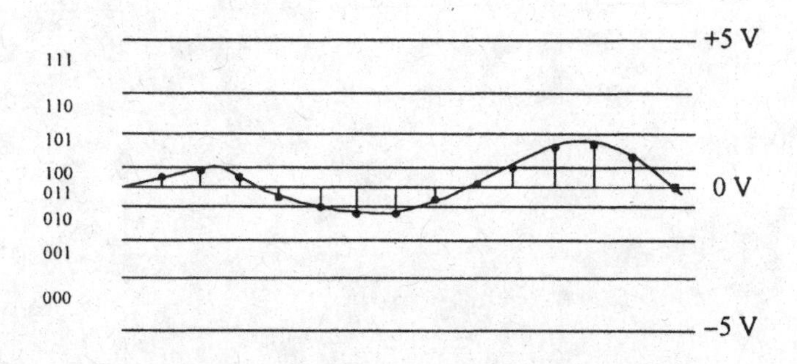

Figure 5.12. Nonuniform quantization.

Figure 5.13 shows a typical companding characteristic. At low signal amplitudes the output signal variation is smaller than the input signal variation. At high levels, just the opposite occurs. The processed signal v_{out} is supplied to a uniform quantizer, sampled, encoded by PCM, and then transmitted. The receiver decodes the input signal, converts the signal to analog form, and then applies the inverse companding characteristic to the signal. A logarithmic curve is often used because the inverse is easy to implement.

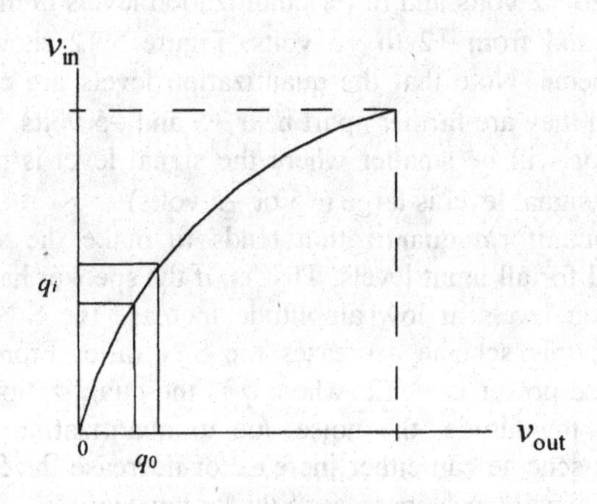

Figure 5.13. Companding.

Chapter 6
Multirate Processing

Multirate processing is important in several applications. One view of wavelet transforms is based on multirate processing. Downsampling with correlation forms the analysis equation (forward transform), and upsampling with convolution forms the synthesis equation (inverse transform). These techniques also apply to the fast Fourier transform (FFT) in Chapter 7.

Many applications use different sampling rates, often in the same system. For example, three different rates are employed in digital audio systems: 32 kHz in broadcasting, 46.1 kHz in digital compact disk (CD), and 48 kHz in digital audio tape (DAT) systems. If one audio amplifier serves all three, we must be able to convert sample rates easily. One way to solve this particular problem would be to use three different D/A conversion units, one for each sample rate. Another way would be to convert two of the signals to the third sample rate before conversion to analog form. This sample rate conversion can be accomplished in the digital domain without using analog signals. How to do this is the subject of this section.

Chapter Goals: After completing this chapter, you should be able to do the following:

- Given $v(n)$ and the downsampling factor D, find and plot the spectrum of the downsampled signal $v_d(n)$.
- Given $v(n)$ and the upsampling factor U, find and plot the spectrum of the upsampled signal $v_u(n)$.
- For signals $h(n)$ and $x(n)$, plot the correlation /downsampling result.
- For signals $h(n)$ and $x(n)$, plot the convolution/upsampling result.

6.1. Downsampling

Downsampling or decimation is the process of decreasing the sampling rate by a factor D, and upsampling or interpolation is the process of increasing the sampling rate by a factor U. These two operations in cascade allow us to change the sampling rate by a factor U/D. First consider downsampling. Figure 6.1 shows downsampling as a system with input $v(n)$ and output $v_d(n)$. Figure 6.2 shows the signal $v(n)$, the intermediate signal $v_a(n)$ where every third sample is selected, and the final decimated signal $v_d(n)$. To derive $v_a(n)$ from $v(n)$, set two out of every three samples to zero. Then throw away these two zero samples to obtain $v_d(n)$. Thus $D = 3$ in this example. In the time domain these signals are related by

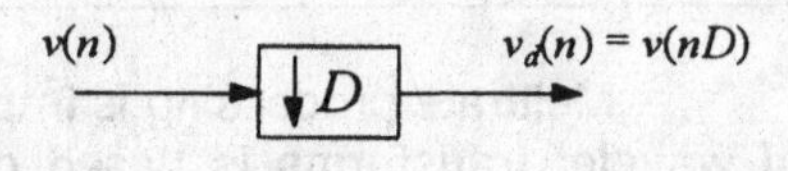

Figure 6.1. Downsampling.

$$v_d(n) = v(nD) = v_a(nD) \tag{6.1}$$

To derive the important relationship between $V(\Omega)$ and $V_d(\Omega)$, define two functions:

$$v_a(n) = \begin{cases} v(n), & n = 0, \pm D, \pm 2D, \ldots \\ 0, & \text{otherwise} \end{cases} \tag{6.2}$$

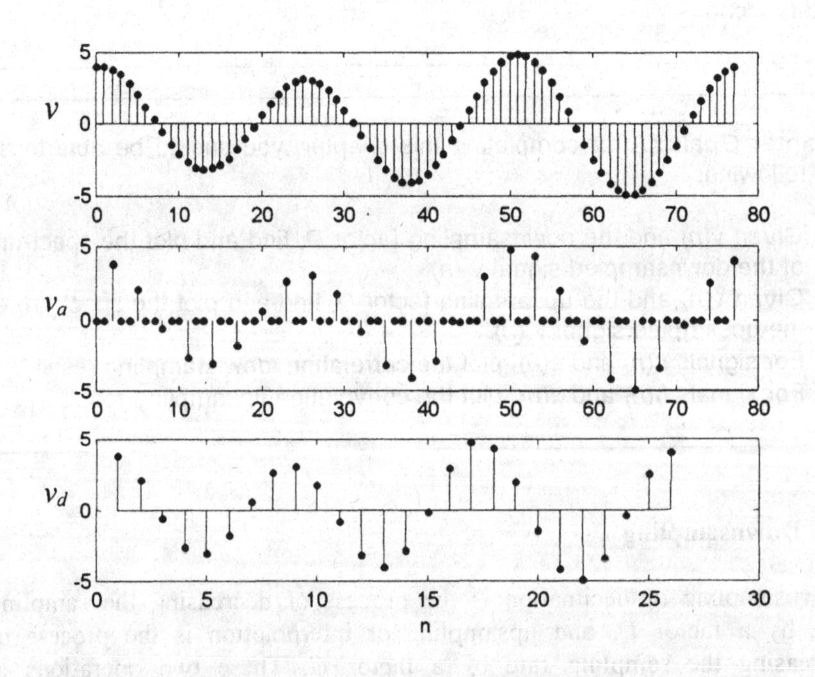

Figure 6.2. Downsampling details.

$$p(n)=\begin{cases}1, & n=0,\ \pm D,\ \pm 2D,\ldots\\ 0, & \text{otherwise}\end{cases} \tag{6.3}$$

Then

$$v_a(n) = p(n)v(n)$$

Figure 6.3 shows the functions $v(n)$, $p(n)$, and $v_a(n)$. Note that multiplying $v(n)$ by $p(n)$ gives the intermediate signal $v_a(n)$. Then delete the zero terms (two out of every three samples) to obtain $v_d(n)$ in Fig. 6.2.

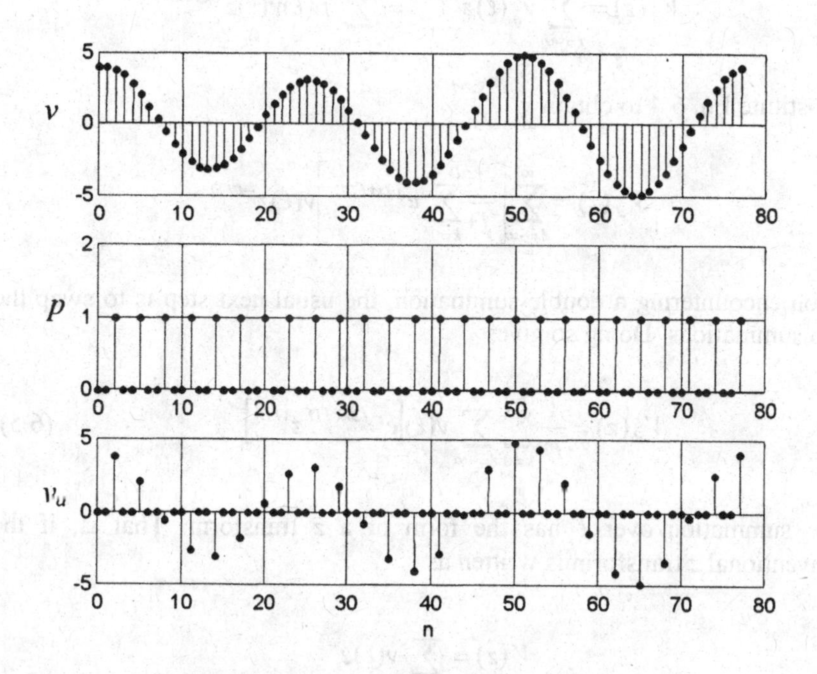

Figure 6.3. The product $v(n)p(n)$ equals $v_a(n)$.

Now we want to represent $p(n)$ by a Fourier series. It will be used later as $p(\ell)$ instead of $p(n)$, so write it as

$$p(\ell) = \frac{1}{D}\sum_{k=0}^{D-1} e^{j2\pi k\ell/D} \tag{6.4}$$

Equations 6.1 through 6.4 can be used to derive a relationship between $V(\Omega)$ and $V_d(\Omega)$. Start with the z transform and then substitute $z = e^{j\Omega}$. The z transform of $v_d(n)$ is given by

$$V_d(z) = \sum_{n=-\infty}^{\infty} v_d(n) z^{-n} = \sum_{n=-\infty}^{\infty} v(nD) z^{-n}$$

Change index of summation. Let $\ell = nD$, or $n = \ell/D$. Then

$$V_d(z) = \sum_{\ell=-\infty}^{\infty} v_a(\ell) z^{-\ell/D} = \sum_{\ell=-\infty}^{\infty} p(\ell) v(\ell) z^{-\ell/D}$$

Substitute Eq. 6.4 to obtain

$$V_d(z) = \sum_{\ell=-\infty}^{\infty} \left[\frac{1}{D} \sum_{k=0}^{D-1} e^{j2\pi k\ell/D} \right] v(\ell) z^{-\ell/D}$$

Upon encountering a double summation, the usual next step is to swap the two summations. Doing so gives

$$V_d(z) = \frac{1}{D} \sum_{k=0}^{D-1} \sum_{\ell=-\infty}^{\infty} v(\ell) \left[e^{-j2\pi k/D} z^{1/D} \right]^{-\ell} \tag{6.5}$$

The summation over ℓ has the form of a z transform. That is, if the conventional z transform is written as

$$V(z) = \sum_{\ell=-\infty}^{\infty} v(\ell) z^{-\ell}$$

The summation in Eq. 6.5 can be written as

$$V(e^{-j2\pi k/D} z^{1/D}) = \sum_{\ell=-\infty}^{\infty} v(\ell) (e^{-j2\pi k/D} z^{1/D})^{-\ell}$$

Therefore, $V_d(z)$ in Eq. 6.5 is given by

$$V_d(z) = \frac{1}{D}\sum_{k=0}^{D-1} V(e^{-j2\pi k/D} z^{1/D}) \tag{6.6}$$

What does this mean? In order to draw a picture, convert this from the z transform to the Fourier transform. Substitute $z = e^{j\Omega}$ to obtain

$$V_d(\Omega) = \frac{1}{D}\sum_{k=0}^{D-1} V\left(e^{j(\Omega - 2\pi k)/D}\right) \tag{6.7a}$$

Or, in more vulgar notation,

$$V_d(\Omega) = \frac{1}{D}\sum_{k=0}^{D-1} V([\Omega - 2\pi k]/D) \tag{6.7b}$$

Recall that $V(\Omega)$ is shorthand notation for the more proper notation $V(e^{j\Omega})$. Equation 6.7 is the relationship we seek. The Fourier transform of a signal that is downsampled by a factor D is given in terms of the original signal V by Eq. 6.7. There are D terms in this equation. If the original signal $v(n)$ has spectrum $V(\Omega) = V(e^{j\Omega})$, the downsampled signal spectrum $V_d(\Omega)$ has D terms in it, where each term is a shifted version of the original. Here are some examples to clarify this equation.

Example 6.1. If $V(\Omega) = V(e^{j\Omega})$ looks like the diagram in Fig. 6.4, plot the spectra $V\left(e^{j(\Omega-2\pi)/D}\right)$ where

(a) $D = 2$.

(b) $D = 3$.

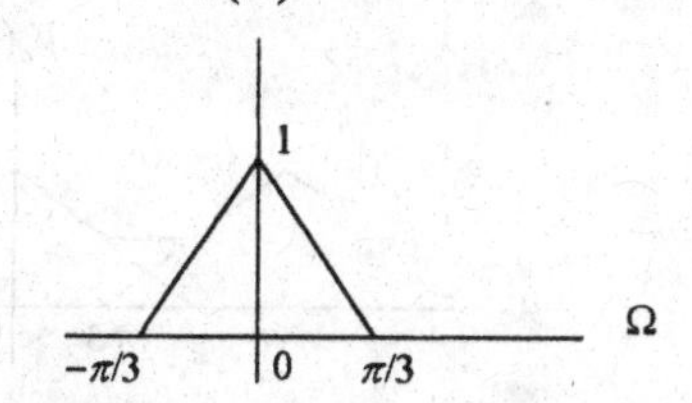

Figure 6.4. $V(\Omega)$.

Solution: For $D = 2$, the spectrum of $V(\Omega/2)$ (before translation) expands by a factor of 2. Consider the unit circle in the Argand diagram. (*Argand diagram* is another name for the complex plane.) The spectrum $V\left(e^{j(\Omega-2\pi)/D}\right)$ is displaced around the unit circle by $2\pi/D = \pi$ units in the counterclockwise direction. When displayed as a conventional Fourier plot (versus Ω), the result is Fig. 6.5a. In a similar manner, when $D = 3$ the displacement is $2\pi/3$ and the function expands by a factor of 3, giving Fig. 6.5b. ■

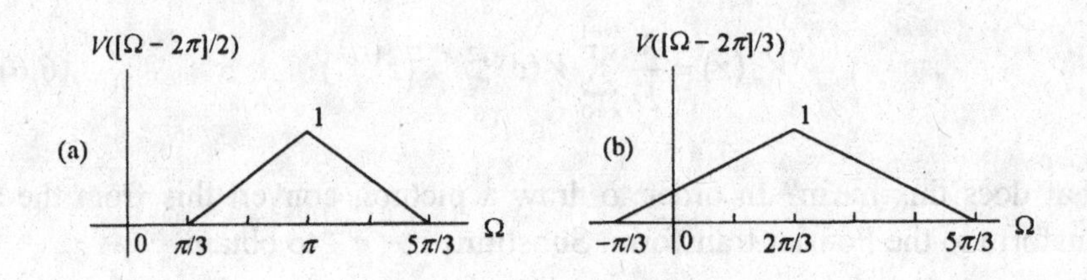

Figure 6.5. Plots of $V\left(e^{j(\Omega-2\pi/D)}\right)$ for $D = 2$ and $D = 3$.

Example 6.2. Plot $V_d(\Omega)$ for $D = 2$ if $V(\Omega)$ is as given in Fig. 6.4.

Solution: With $D = 2$, Eq. 6.7a gives

$$V_d(\Omega) = \tfrac{1}{2}\left\{V(e^{j\Omega/2}) + V(e^{j(\Omega-2\pi)/2}\right)$$

There are two terms, one centered at 0 and the other centered at π, each with amplitude ½, as shown in Fig. 6.6. However,in discrete time, this plot is periodic with period 2π. There is considerable spectral overlap (aliasing), indicating the need for filtering to limit the spectral width of $V(\Omega)$. ■

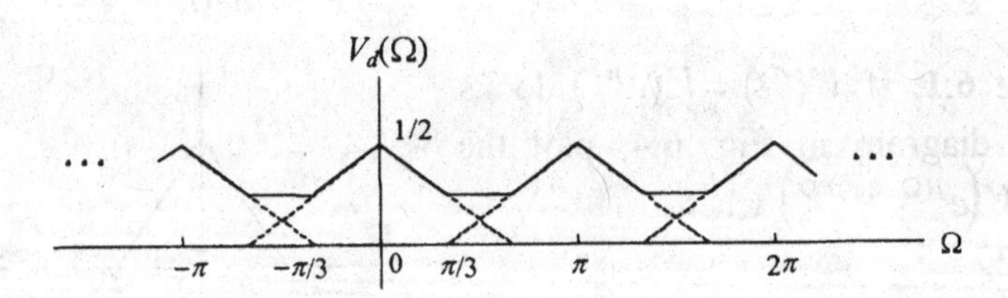

Figure 6.6. Solution to Example 6.2.

For another look at this process, Fig. 6.7 continues our discussion for the case $D = 3$. When $D = 3$, Eq. 6.7 gives

$$V_d(\Omega) = \tfrac{1}{3}\left\{V\left(e^{j\Omega/3}\right) + V\left(e^{j(\Omega-2\pi)/3}\right) + V\left(e^{j(\Omega-4\pi)/3}\right)\right\}$$

Note that $V_a(\Omega)$ in Fig. 6.7b has three terms in it for every term in the original signal $V(\Omega)$. The bandwidth of $V(\Omega)$ is $\pi/3$, so there is no overlap (no aliasing) in $V_a(\Omega)$, but if the bandwidth of $V(\Omega)$ had been any larger, aliasing would have occurred. In order to derive $V_d(\Omega)$ from $V(\Omega)$ of Fig. 6.7, note that

$$v_d(n) = v(nD) = v_a(nD) \qquad \text{Repeated (6.1)}$$

Let the notation $v_\ell(n)$ stand for the function

$$v_\ell(n) = \begin{cases} v\left(\dfrac{n}{\ell}\right), & \text{for } n \text{ a multiple of } \ell \\ 0, & \text{otherwise} \end{cases}$$

The time-expansion property of the DTFT is given by

$$v_\ell(n) \leftrightarrow V(\ell\Omega)$$

In Fig. 6.7 the signal $v_d(n)$ plays the role of $v(n)$, and $v_a(n)$ plays the role of $v_\ell(n)$ in the time-expansion property. Therefore, the spectrum of $v_a(n)$ is related to the spectrum of $v_d(n)$ by the relationship

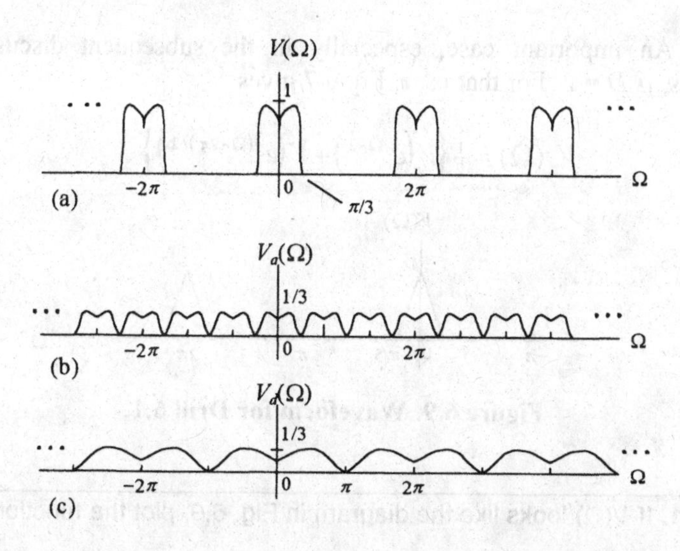

Figure 6.7. Closer look at downsampling.

$$v_a(n) \leftrightarrow V_a(\Omega) = V_d(D\Omega)$$

Conversely,

$$V_d(n) \leftrightarrow V_d(\Omega) = V_a\left(\frac{\Omega}{D}\right)$$

This relationship is evident if you compare parts (b) and (c) of Fig. 6.7.

We can avoid aliasing by placing a low-pass filter before the downsampling operation. The bandwidth of this filter is determined by the factor D independent of the signal bandwidth. That is, to avoid aliasing, filter the signal so that its maximum frequency is $\Omega_m = \pi/D$, as you can see from Fig. 6.7, where $D = 3$. Of course, this introduces distortion in the signal if the analog signal bandwidth is more than $\pi r_s/D$ (r_s is the sampling rate). Figure 6.8 shows this filter-downsampling combination.

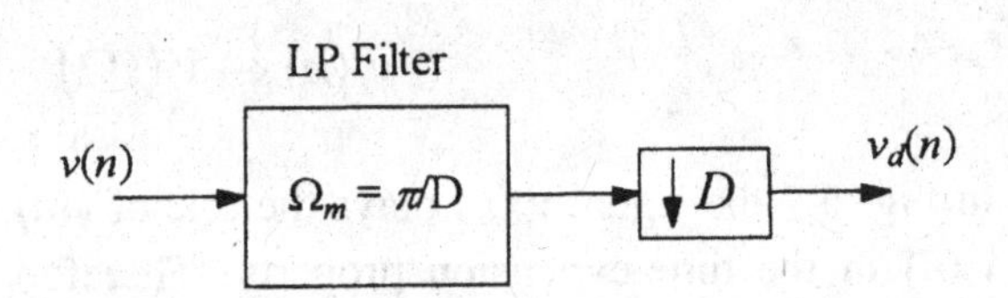

Figure 6.8. Filtering and downsampling.

An important case, especially in the subsequent discussion of wavelets, is $D = 2$. For that case, Eq. 6.7 gives

$$V_d(\Omega) = \tfrac{1}{2}\left\{V\left(e^{j\Omega/2}\right) + V\left(e^{j(\Omega-2\pi)/2}\right)\right\} \tag{6.8}$$

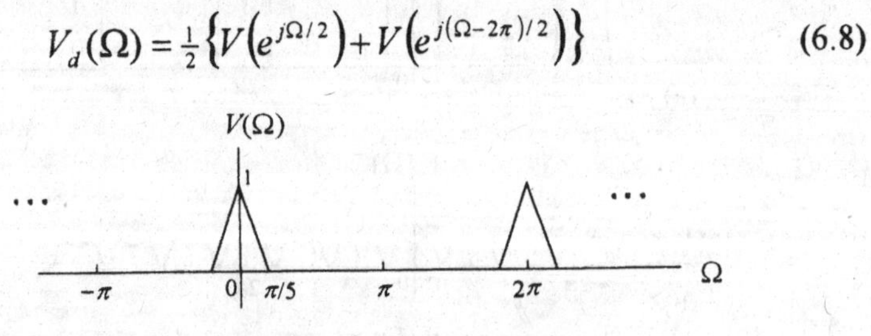

Figure 6.9. Waveform for Drill 6.1.

Drill 6.1. If $V(\Omega)$ looks like the diagram in Fig. 6.9, plot the function

$$V_d(\Omega) = \tfrac{1}{2}\left\{V\left(e^{j\Omega/2}\right) + V\left(e^{j(\Omega-2\pi)/2}\right)\right\}$$

Answer: Figure 6.10.

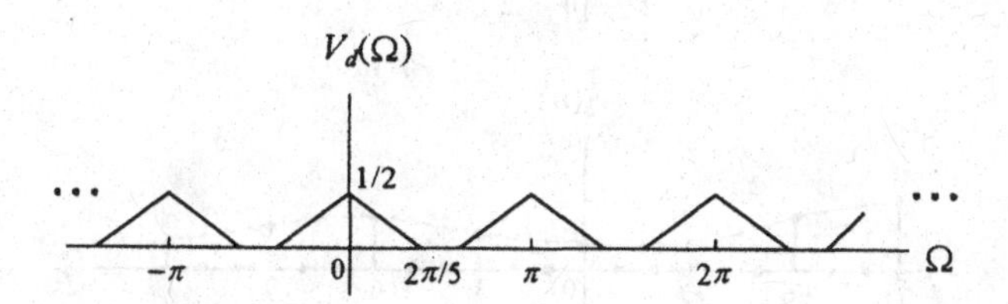

Figure 6.10. Answer for Drill 6.1.

6.2. Upsampling

Figure 6.11 shows the process of upsampling (interpolation) by a factor U, along with a low-pass filter. Upsampling is accomplished by inserting U – 1 zero samples between each sample of $v(n)$ to obtain $v_b(n)$, and then filtering to obtain $v_u(n)$, as shown in Fig. 6.12. Upsampling expands $v(n)$ in the time domain, therefore shrinking it in the frequency domain.

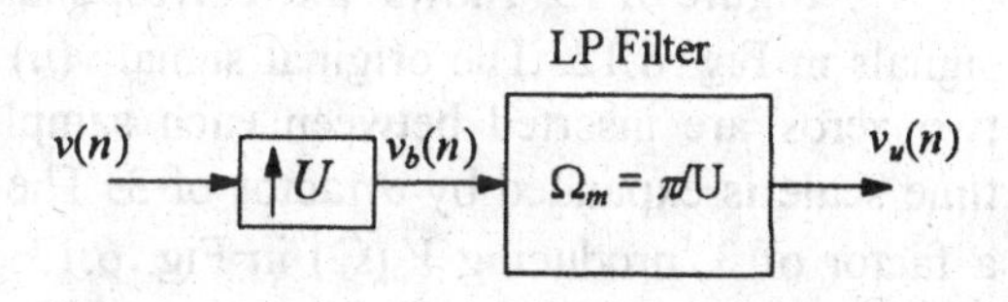

Figure 6.11. Up sampling and filtering.

To express $v_b(n)$ in terms of $v(n)$, write

$$v_b(n) = \begin{cases} v(n/U), & n = 0, \pm U, \pm 2U, \ldots \\ 0, & \text{otherwise} \end{cases}$$

This is $v_\ell(n)$ in the time-expansion property with $\ell = U$. This gives the Fourier transform as

$$V_b(\Omega) = V(U\Omega) = \sum_{n=-\infty}^{\infty} v(n)e^{-jnU\Omega}$$

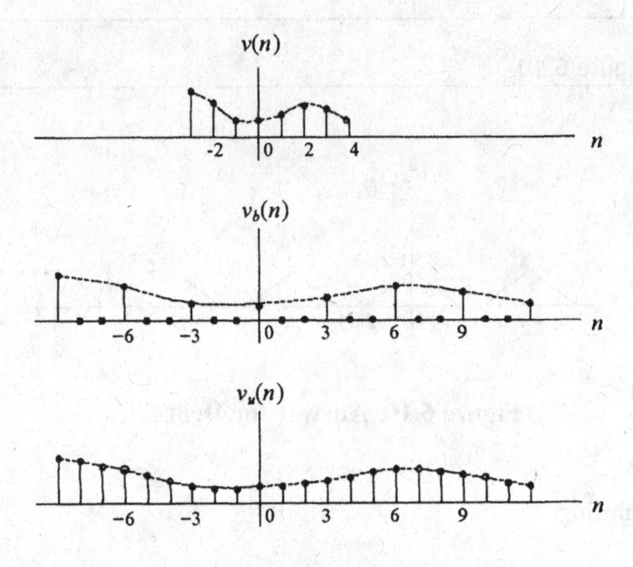

Figure 6.12. Upsampling.

Figure 6.13 shows the corresponding frequency spectra for the signals in Fig. 6.12. The original signal $v(n)$ has the spectrum $V(\Omega)$. When two zeros are inserted between each sample of $v(n)$ to obtain $v_b(n)$, the time scale is expanded by a factor of 3. The frequency scale is reduced by a factor of 3, producing $V_b(\Omega)$ in Fig. 6.13. The low-pass filter eliminates the "extra" spectral components to produce $V_u(\Omega)$. This time the filter bandwidth should be no more than π/U.

6.3. Fractional Rate Change

The process of upsampling by U and downsampling by D can be combined to change the rate by a factor U/D.

$$r_s' = \frac{U}{D} r_s \tag{6.9}$$

Figure 6.14 shows upsampling first followed by downsampling, which places the two low pass-filters in cascade. Since both are low-pass filters, they can be replaced by a single filter with bandwidth π/M, where π/M is the minimum between π/U and π/D. The process of changing sample rate consists of upsampling, low-pass filtering, followed by downsampling.

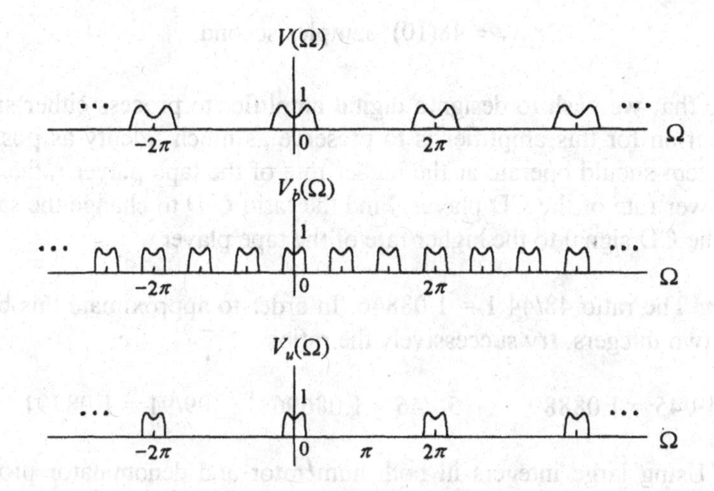

Figure 6.13. Spectrum for the signals in Fig. 6.12.

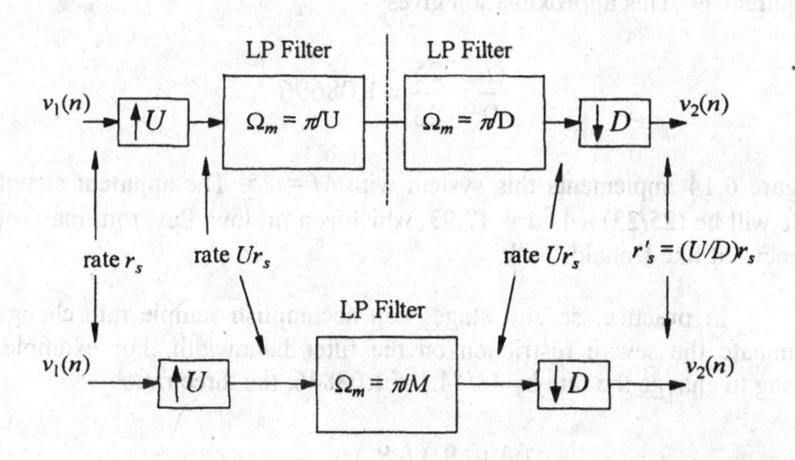

Figure 6.14. Replacing the two filters by one filter.

Example 6.3. Many audio systems are designed to play both CD and digital tapes. These devices operate at different sampling rates.

$$r_{\text{CD}} = 44.1(10)^3 \text{ samples/second}$$

$$r_{\text{DAT}} = 48(10)^3 \text{ samples/second}$$

Suppose that we wish to design a digital amplifier to process either signal. One criterion for this amplifier is to preserve as much fidelity as possible. This system should operate at the higher rate of the tape player rather than at the lower rate of the CD player. Find the ratio U/D to change the sample rate of the CD signal to the higher rate of the tape player.

Solution: The ratio 48/44.1 = 1.08846. In order to approximate this by the ratio of two integers, try successively the ratios

$$49/45 = 1.08889 \qquad 50/46 = 1.08696 \qquad 99/91 = 1.08791$$

Using large integers in both numerator and denominator provides a closer approximation to the correct ratio 1.08846. Even the smaller ratio 50/46 = 25/23 has numbers too large for our application. To use these numbers the sampling rate must increase by 25, and then decrease by 23. This requires a low-pass filter of bandwidth $\pi/25$, a rather severe requirement. This approximation gives

$$\frac{U}{D} = \frac{25}{23} = 1.08696$$

Figure 6.14 implements this system with $M = 25$. The apparent sampling rate will be $(25/23) \times 44.1 = 47.93$, which is a bit low. Pavarotti may sound somewhat like Donald Duck.

■

In practice, several stages can accomplish sample rate change to eliminate the severe restriction on the filter bandwidth. For example, in trying to change the rate by 48/44.1 = 1.08846, the three ratios

$$\left(\frac{4}{3}\right)\left(\frac{9}{10}\right)\left(\frac{9}{10}\right) = 1.0800$$

are reasonably close. More trials of different combinations of small numbers will yield a closer approximation. Improvement occurs if we realize that 9/10 = (3/5)(3/2). Therefore, a total of five stages implements a sample rate change of 1.08, with the largest value of M in Fig. 6.14 equal to 5. How much it costs versus how much error one is willing to tolerate governs the design.

Downsampling and upsampling are important in understanding wavelets. The following three examples use MATLAB to explore the frequency content of related signals.

Example 6.4. Find and compare the frequency content of the four signals in Fig. 6.15.

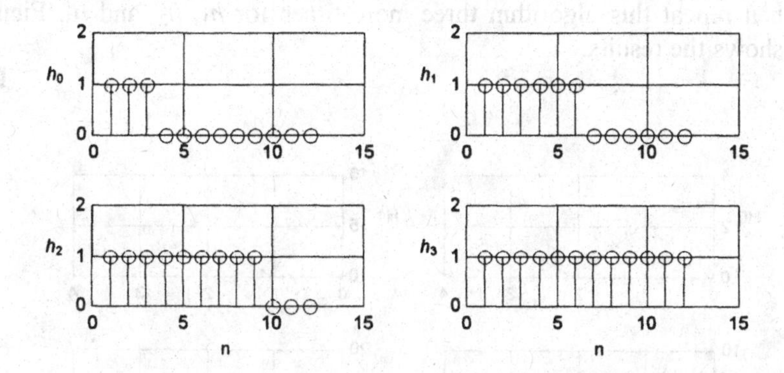

Figure 6.15. The four signals in Example 6.4.

Solution: The first part of the following MATLAB program generates the signals h_0, h_1, h_2, and h_3. The next part calculates the Fourier transform of h_0 for 100 values of Ω in the interval $(0,\pi)$, and plots the magnitude.

```
N=12;
M=3;
h0=[ones(1,M)  zeros(1,N-M)];
M=6;
h1=[ones(1,M)  zeros(1,N-M)];
M=9;
h2=[ones(1,M)  zeros(1,N-M)];
M=12;
h3=ones(1,M);
```

Calculate and plot the transform of h_0. This is the first graph in Fig. 6.16, labeled H0.

```
w=linspace(0,pi);
s=zeros(size(w));
for n=1:N
```

```
  k=n-1;
  s=s+h0(n)*exp(-j*w*k);
end
H0=abs(s);
subplot(4,2,1)
plot(w,H0)
ylabel('H0')
```

We then repeat this algorithm three more times for h_1, h_2, and h_3. Figure 6.16 shows the results.

■

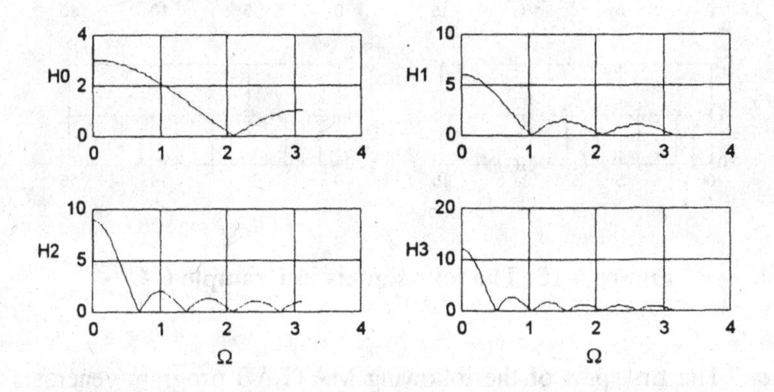

Figure 6.16. Frequency content of the signals in Fig. 6.15.

Example 6.5. *Decimation*. This example demonstrates the effect of replacing samples by zeros. Figure 6.17 shows the original signal, consisting of 12 samples, followed by the removal of samples in the succeeding diagrams. (We used 12 samples because 12 is divisible by 2, 3, and 6.) Figure 6.18 shows the frequency content of these four signals. The MATLAB programs that produced these figures are similar to those of Example 6.4.

■

Example 6.6. *Interpolation*. This example demonstrates the effect of inserting zeros between samples, which is the intermediate step in interpolation. Figure 6.19 shows a signal consisting of three samples. We then insert one, two, and three zero-valued samples between each sample. Figure 6.20 shows the frequency content of the corresponding signals.

■

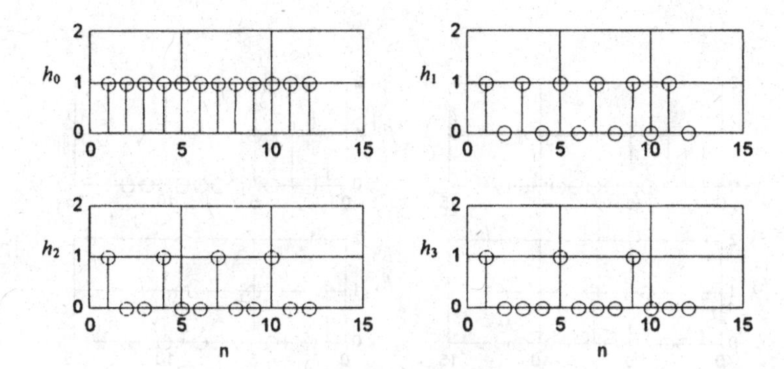

Figure 6.17. Signals for Example 6.5.

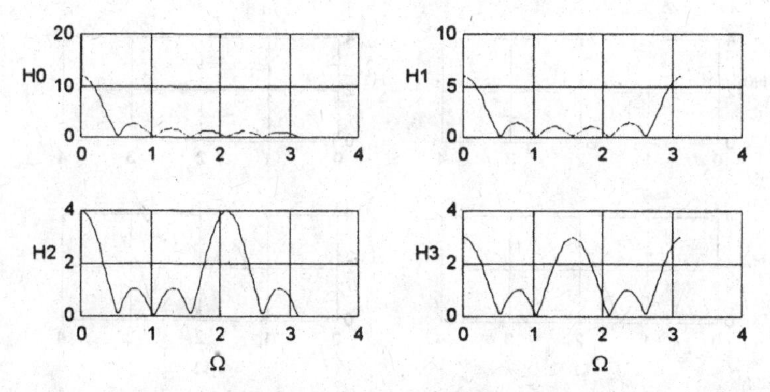

Figure 6.18. Frequency content of the signals in Fig. 6.17.

Note the similarities and differences in the frequency content of these signals. In Figs. 6.15 and 16, the signal length in the time domain determines the width of the frequency representation, illustrating time-frequency duality. In Fig. 6.17 we start with a signal of length 12 and interleave zero samples, producing the spectra in Fig. 6.18. Figures 6.19 and 6.20 are similar, the only difference being that we use only three samples throughout. From these diagrams we see that Figs. 6.18 and 6.20 differ only because of the different number of samples used. These diagrams establish a base to use in comparing the effects of adding zero-valued samples.

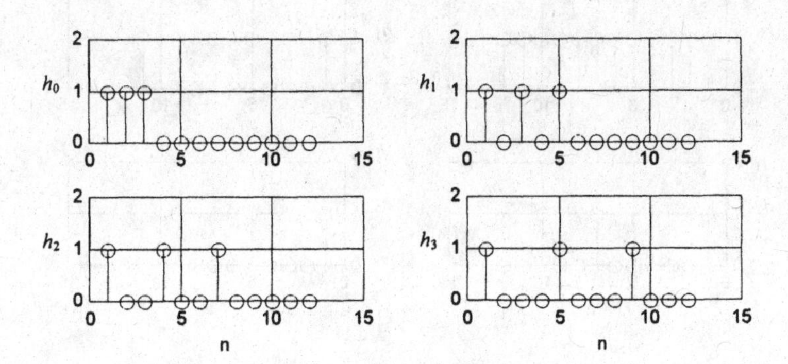

Figure 6.19. Interpolation.

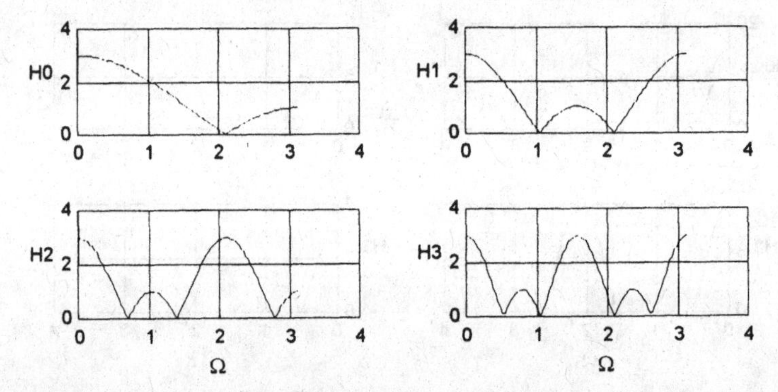

Figure 6.20. Frequency content of the interpolated signals.

6.4. Downsampling and Correlation

Like anything else, correlation and convolution with downsampled (or upsampled) signals is easy once you know how. Our purpose in this section is to show you how. We begin with correlation. Figure 6.21 shows correlation followed by downsampling. Recall that correlation with $h(n)$ is equivalent to convolution with $h(-n)$. For that reason, correlation is depicted as filtering with $h(-n)$. The correlation operation is given by

$$v(n) = \sum_k x(k)h(k-n) \tag{6.10}$$

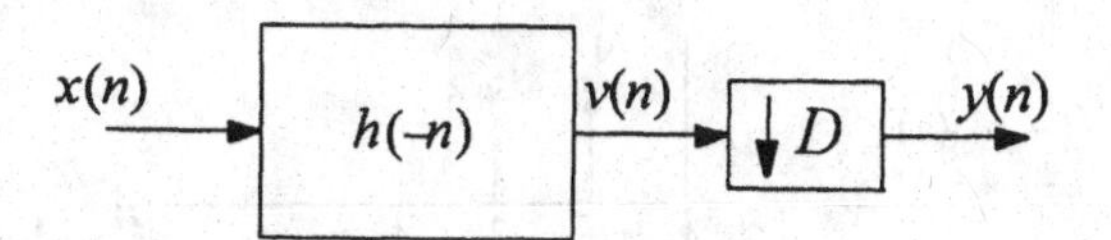

Figure 6.21. Correlation followed by downsampling.

Example 6.7. Figure 6.22 shows $h(n)$ and $x(n)$. Use Eq. 6.10 to obtain $v(n)$, then downsample by 2 to obtain $y(n)$.

Solution: Figure 6.23a shows the result of correlating $x(n)$ with $h(n)$. Downsampling by 2 selects the even-numbered components of $v(n)$ to produce $y(n)$ in Fig. 6.23b.

■

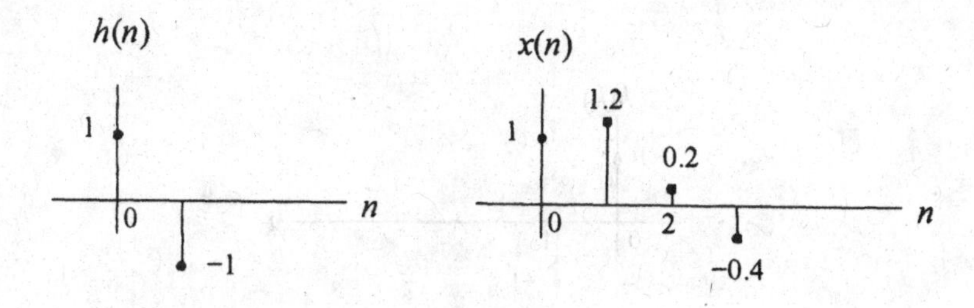

Figure 6.22. The impulse response $h(n)$ and input $x(n)$.

One purpose in this example is to demonstrate that correlation followed by downsampling can be combined into one equation:

$$y(n) = \sum_k x(k)h(k - Dn) \tag{6.11}$$

This equation describes the operation in Fig. 6.21. Redoing the preceding example using Eq. 6.11 in Fig. 6.24, you can see that the sum over k of the product $h(k)x(k)$ produces $y(0) = -0.2$, and similarly $h(k-2)x(k)$ produces $y(1) = 0.6$.

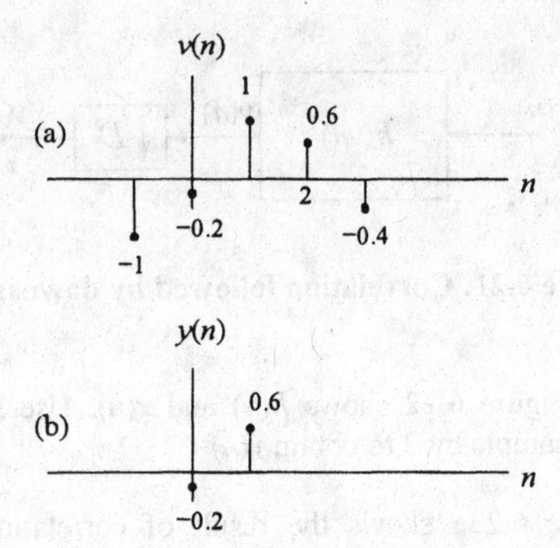

Figure 6.23. Producing $y(n)$.

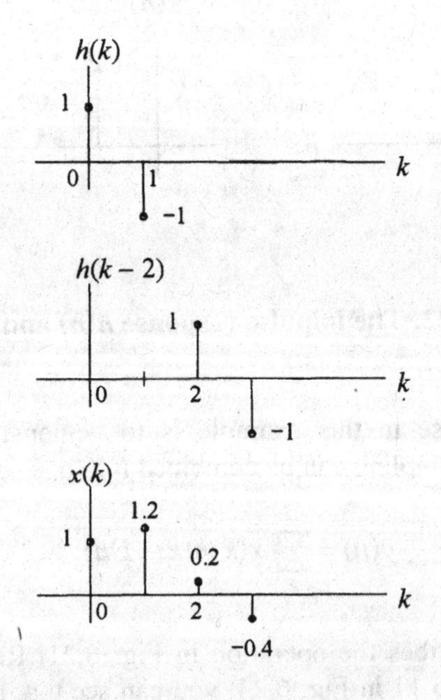

Figure 6.24. Illustrating Eq. 6.11.

Two formulas that will be derived in Chapter 10 relate wavelet coefficients $c_j(k)$ and $d_j(k)$ as follows:

$$c_j(k) = \sum_m h_0(m-2k)c_{j+1}(m) \qquad k \geq 0 \tag{6.12}$$

$$d_j(k) = \sum_m h_1(m-2k)c_{j+1}(m) \qquad k \geq 0 \tag{6.13}$$

These equations say that for given sequences h_0 and h_1, we can start with a given sequence c_{j+1} and derive two other sequences, c_j and d_j. Although this has no significance now, it will assume importance in our study of wavelets. Here is an example to demonstrate the use of these formulas.

Example 6.8. Derive the c_1, d_1, c_0, and d_0 sequences if c_2, h_0, and h_1 are given by

$$c_2 = [0.8183 \quad 0.8183 \quad 0.1817 \quad 0.1817]$$

$$h_0 = \left[\tfrac{1}{\sqrt{2}} \quad \tfrac{1}{\sqrt{2}}\right] \qquad h_1 = \left[\tfrac{1}{\sqrt{2}} \quad -\tfrac{1}{\sqrt{2}}\right]$$

Solution: Let us first derive the c_1 sequence from Eq. 6.12. Figure 6.25 shows $c_2(m)$, $h_0(m)$, and $h_0(m-2)$. Setting $k = 0$ in Eq. 6.12, multiplying, and then summing gives

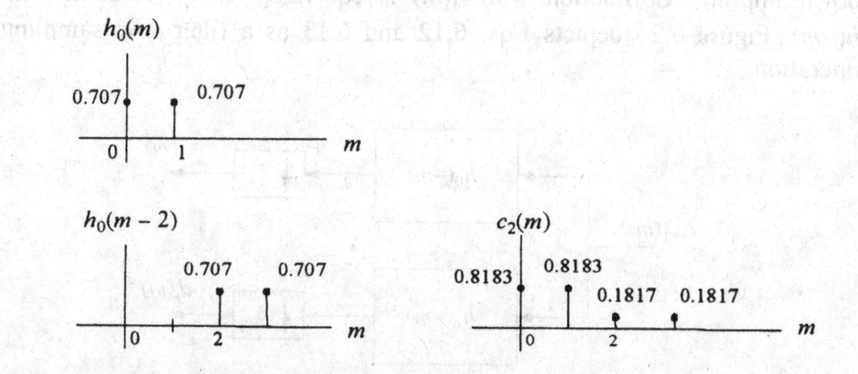

Figure 6.25. The functions $h_0(m)$, $h_0(m-2)$, and $c_2(m)$,

$$c_1(0) = \sum_m h_0(m)c_2(m) = \frac{1}{\sqrt{2}}[0.8183 + 0.8183] = 1.1573$$

Next, with $k = 1$ we get

$$c_1(1) = \sum_m h_0(m-2)c_2(m) = \frac{1}{\sqrt{2}}[0.1817 + 0.1817] = 0.2570$$

In a similar manner, the d_1 sequence is given by

$$d_1(0) = \sum_m h_1(m)c_2(m) = \frac{1}{\sqrt{2}}[0.8183 - 0.8183] = 0$$

$$d_1(1) = \sum_m h_1(m-2)c_2(m) = \frac{1}{\sqrt{2}}[0.1817 - 0.1817] = 0$$

Finally,

$$c_0(0) = \sum_m h_0(m)c_1(m) = \frac{1}{\sqrt{2}}[1.1573 + 0.2570] = 1$$

$$d_0(0) = \sum_m h_1(m)c_1(m) = \frac{1}{\sqrt{2}}[1.1573 - 0.2570] = 0.6366$$

■

Notice that this example started with a sequence of length 4, $c_2(m)$. After correlation and downsampling, the resulting sequence $c_1(m)$ has length 2, and one more operation reduces the sequence to length 1. Also, notice that the process of deriving c_j from c_{j+1} is equivalent to filtering and downsampling. Correlation with $h(m)$ is equivalent to convolution with $h(-m)$. Figure 6.26 depicts Eqs. 6.12 and 6.13 as a filter-downsampling operation.

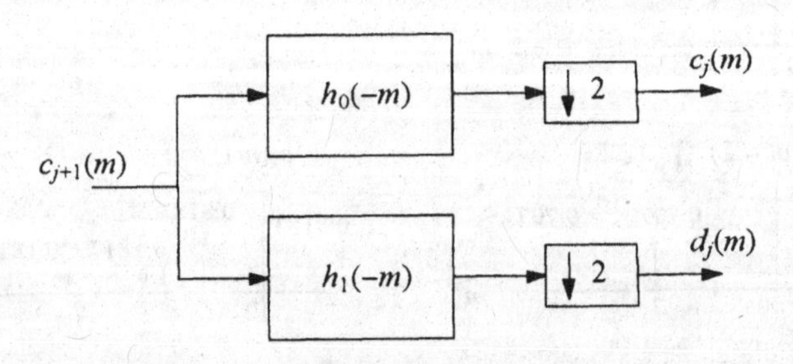

Figure 6.26. The filter-down sample operations of Eqs. 6.12 and 6.13.

Example 6.9. Derive the c_2 and d_2 coefficients if c_3, h_0, and h_1 are given by

$$c_3 = [4,\ 5,\ -4,\ 12,\ -5,\ 11,\ -2,\ 1]$$
$$h_0 = [3,\ 5,\ 1,\ -0.5]$$
$$h_1 = [-0.5,\ -1,\ 5,\ -3]$$

Solution: Correlating h_0 with c_3 gives

$$h_0 \otimes c_3 = [\underset{-3}{-2}\quad \underset{-2}{1.5}\quad \underset{-1}{27}\quad \underset{0}{27}\quad \underset{1}{9.5}\quad \underset{2}{37.5}\quad \underset{3}{23}\quad \underset{4}{37.5}\quad \underset{5}{24}\quad \underset{6}{-1}\quad \underset{7}{3}]$$

The small index numbers below the correlation values represent time. If $h_0(n)$ and $c_3(n)$ start at $n = 0$, their correlation starts an $n = -3$ and ends at $n = 7$. To determine c_2, use the even correlation values starting at $n = 0$. This gives

$$c_2 = [27\quad 37.5\quad 37.5\quad -1]$$

In a similar manner, correlating h_1 with c_3 and downsampling gives

$$h_1 \otimes c_3 = [-12\quad 5\quad 33\quad -63\quad 76.5\quad -68\quad 60\quad -21.5\quad 1.5\quad 0\quad -0.5]$$

Selecting the even-indexed values, starting at $n = 0$, gives

$$d_2 = [-63\quad -68\quad -21.5\quad 0]$$

■

Drill 6.2. Given

$$h_0 = \{1,\ 1,\ -1,\ -1\}$$
$$h_1 = \{-1,\ 1,\ 1,\ -1\}$$
$$c_3 = \{1,\ 2,\ 1,\ 0,\ -1,\ 1,\ 0,\ 1\}$$

find c_2 and d_2.

Answer: $c_2 = \{2,\ 1,\ -1,\ 1\}$, $d_2 = \{2,\ -3,\ 1,\ 1\}$

6.5. Upsampling and Convolution

Figure 6.27 shows the operation of upsampling by U, followed by filtering. First upsample $x(n)$ to produce $v(n)$, followed by convolution with $h(n)$ to produce $y(n)$. Before going further, let us explore the operations in Fig. 6.27 with an example.

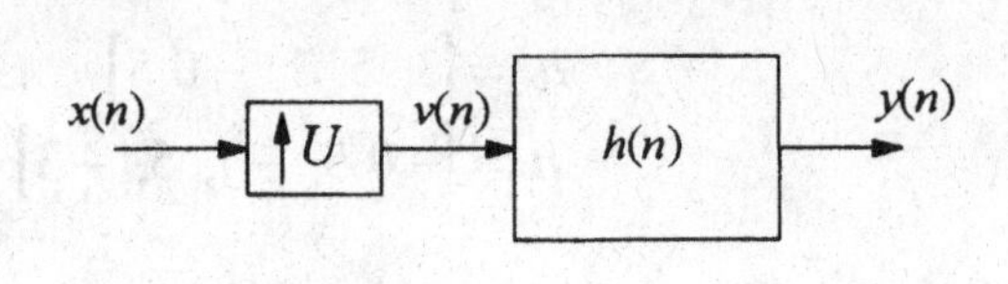

Figure 6.27. Upsampling and convolution.

Example 6.10. Figure 6.28 shows $x(n)$ and $h(n)$. Perform the operations in Fig. 6.27 with $U = 2$ to obtain $y(n)$.

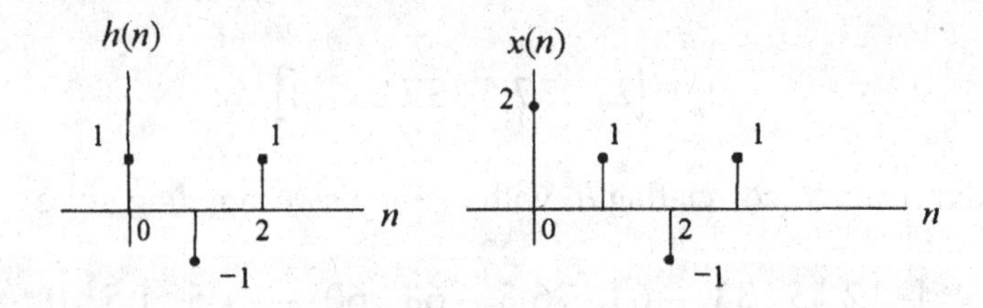

Figure 6.28. The functions in Example 6.10.

Solution: Figure 6.29a shows the result of upsampling $x(k)$ to obtain $v(k)$. Insert one zero sample between each value of $x(k)$ to upsample by 2. Convolve $v(k)$ with $h(k)$ by the formula

$$y(n) = \sum_k h(n-k)v(k)$$

to obtain $y(n)$ in Fig. 6.29c.

■

This example illustrates the two-step process of upsampling by a factor U followed by convolution with $h(n)$. For reasons explained later (in connection with wavelets), we would like to combine these two operations into one expression. The formula that describes both the operations of upsampling and convolution is given by

$$y(n) = \sum_{k} h(n - Uk)x(k) = \sum_{\lambda} h(\lambda)x\left(\frac{n-\lambda}{U}\right) \tag{6.14}$$

where the second summation is derived from the first by a change of variable, $\lambda = n - Uk$ or $k = (n - \lambda)/U$. As the next example illustrates, the combination of these two operations into one expression leads to complexity.

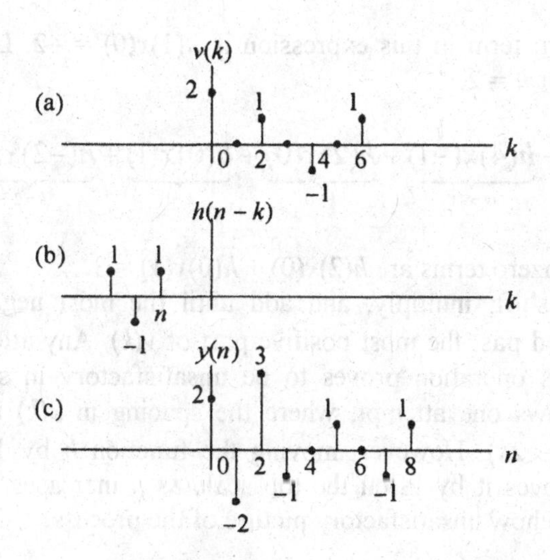

Figure 6.29. Upsampling followed by convolution.

Example 6.11. Use Eq. 6.14 for the functions in Fig. 6.28 with $U = 2$ to obtain $y(n)$.

Solution: Here is where a real problem occurs if we try to draw pictures that illuminate the process in Eq. 6.14. It is best to take a different (simpler) approach before trying to draw an appropriate picture. Evaluate the left side of Eq. 6.14 one step at a time. First let $n = 0$. Then Eq. 6.14 gives

$$y(0) = \sum_k h(-Uk)x(k)$$
$$= \cdots + \underbrace{h(2)x(-1)}_{k=-1} + \underbrace{h(0)x(0)}_{k=0} + \underbrace{h(-2)x(1)}_{k=1} + \underbrace{h(-4)x(2)}_{k=2} + \cdots$$

The only nonzero term in this expression is $h(0)x(0) = 2$. similarly, if $n = 1$ in Eq. 6.14, we get

$$y(1) = \cdots + \underbrace{h(3)x(-1)}_{k=-1} + \underbrace{h(1)x(0)}_{k=0} + \underbrace{h(-1)x(1)}_{k=1} + \underbrace{h(-3)x(2)}_{k=2} + \cdots$$

The only nonzero term in this expression is $h(1)x(0) = -2$. Let us do one more of these, for $n = 2$.

$$y(2) = \cdots + \underbrace{h(4)x(-1)}_{k=-1} + \underbrace{h(2)x(0)}_{k=0} + \underbrace{h(0)x(1)}_{k=1} + \underbrace{h(-2)x(2)}_{k=2} + \cdots$$

Here the two nonzero terms are $h(2)x(0) + h(0)x(1) = 3$.

Simply shift, multiply, and add until the most negative part of $h(n-k)$ is shifted past the most positive part of $v(k)$. Any attempt to draw a picture of this operation proves to be unsatisfactory in some respect. Figure 6.30 shows one attempt, where the spacing in $x(k)$ is double the spacing in $h(n-2k)$. However, moving the function h by 1 step on the bottom scale moves it by ½ on the top scale as n increases. This gives a correct, but somehow unsatisfactory picture of the process.

■

Drill 6.3. Given the two sequences

$$h(n) = \{1,\ 1,\ -1,\ -1\}$$
$$x(n) = \{1,\ 2,\ 1,\ 0,\ -1,\ 1\}$$

upsample $x(n)$ by $U = 3$ and convolve with $h(n)$ to obtain $y(n)$.

Answer:

$y(n) = [1, 1, -1, 1, 2, -2, -1, 1, -1, -1, 0, 0, -1, -1, 1, 2, 1, -1, -1]$

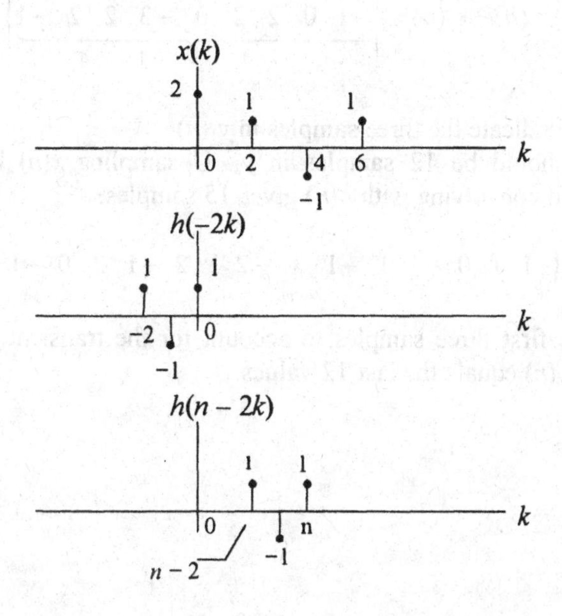

Figure 6.30. Attempt to illustrate Eq. 6.16.

Example 6.12. Correlation followed by downsampling by 2, and upsampling by 2 followed by convolution, are two processes that will be used in connection with wavelets. Note the order, correlation first, followed by downsampling, or upsampling first, followed by convolution. For the signal $x(n)$ and filter $h(n)$ given by

$$x(n) = \{-1 \quad 0 \quad 1 \quad 3 \quad 1 \quad -1\}$$
$$h(n) = \{1 \quad -1 \quad 0 \quad 1\}$$

(a) Correlate and downsample to obtain $y_1(n)$. How many samples are there in $y_1(n)$?
(b) Upsample and convolve to obtain $y_2(n)$. How many samples in $y_2(n)$?

Solution: (a) Since there are six samples in x, there should be three samples in y_1. Correlating x with h results in

$$(h \otimes x)(n) = \left\{ \underbrace{-1 \;\; 0 \;\; 2 \;\; \underset{\uparrow}{2} \;\; 0 \;\; \underset{\uparrow}{-3} \;\; \underset{\uparrow}{2} \;\; 2 \;\; -1} \right\}$$

The arrows indicate the three samples in $y_1(n)$.

(b) There should be 12 samples in y_2. Upsampling $x(n)$ by 2 gives 12 samples, and convolving with $h(n)$ gives 15 samples.

$$\{-1 \;\; 1 \;\; 0 \;\; -1 \;\; 1 \;\; -1 \;\; 3 \;\; -2 \;\; 1 \;\; 2 \;\; -1 \;\; 2 \;\; 0 \;\; -1 \;\; 0\}$$

Discard the first three samples to account for the transient portion in this signal, so $y_2(n)$ equals the last 12 values.

■

Chapter 7
Fast Fourier Transform

The fast Fourier transform (FFT) performs the same operation on a signal as the discrete time Fourier series (DTFS). They are the same mathematically, the difference being in the way they go about it. This chapter presents three views of the FFT. First, the DTFS can be viewed as a linear transformation between two finite-dimensional vector spaces. Hence, there is a matrix of transformation. The FFT is a particular decomposition of this matrix operation that leads to fewer multiplication and addition operations. The second view of the FFT, using signal flow graphs, is closely related to the first. Signal flow graphs are a method of graphically representing algebraic equations. The third view of the FFT uses the concept of downsampling in a divide-and-conquer paradigm. Of course, these are all just different ways of looking at the same thing, so they are equivalent. No one view is superior to the others, for all may prove to be useful in different applications.

Chapter Goals: After completing this chapter, you should be able to do the following:

- Find the matrix of transformation for a given *N*-point FFT.
- Calculate the DTFS using matrix decomposition.
- Calculate the DTFS using downsampling.

7.1. Discrete-Time Fourier Series

The DTFS is a transformation between finite-dimensional vector spaces. The domain is the set of all discrete-time signals of length *N*, and the codomain is the set of all discrete-frequency signals of length *N*. That is,

$$\text{DTFS}: V \to V : v(n) \mapsto \sum_{n=0}^{N-1} v(n)e^{-j2\pi nk/N} \tag{7.1}$$

where V is a set of complex-valued sequences of length *N*. This notation reminds us that the DTFS is a transformation between finite-dimensional vector spaces. Equation 7.1 is read, "the DTFS is a map (or function) from V to V such that $v(n)$ is transformed into $\sum_{n=0}^{N-1} v(n)e^{-j2\pi nk/N}$." The techniques from Chapter 4 apply to the DTFS, meaning that a matrix represents the DTFS. The algorithm on page 70, culminating in Eq. 4.3, provides the method of finding this matrix.

1. First select bases $\alpha = \{\alpha_i\}$ and $\beta = \{\beta_i\}$ for the domain and codomain, respectively.
2. Apply the DTFS to each basis vector α_i to obtain $V_i(k)$.
3. Find $[V_i(k)]_\beta$, the coordinates of $V_i(k)$ with respect to the codomain basis β.
4. The matrix of transformation is composed of the column vectors $[V_i(k)]_\beta$.

Let us apply this to sequences of length 4. The matrix of transformation will be a 4×4 matrix W.

Step 1: Select as bases

$$\alpha = \beta = \left\{ \begin{bmatrix} 1 \\ 0 \\ 0 \\ 0 \end{bmatrix} \begin{bmatrix} 0 \\ 1 \\ 0 \\ 0 \end{bmatrix} \begin{bmatrix} 0 \\ 0 \\ 1 \\ 0 \end{bmatrix} \begin{bmatrix} 0 \\ 0 \\ 0 \\ 1 \end{bmatrix} \right\} \tag{7.2}$$

Step 2: Transform each basis vector. (Apply Eq. 7.1.) This gives

$$\alpha_1 = \begin{bmatrix} 1 \\ 0 \\ 0 \\ 0 \end{bmatrix} \leftrightarrow \begin{bmatrix} 1 \\ 1 \\ 1 \\ 1 \end{bmatrix} = V_1(k) \qquad \alpha_2 = \begin{bmatrix} 0 \\ 1 \\ 0 \\ 0 \end{bmatrix} \leftrightarrow \begin{bmatrix} 1 \\ -j \\ -1 \\ j \end{bmatrix} = V_2(k)$$

$$\alpha_3 = \begin{bmatrix} 0 \\ 0 \\ 1 \\ 0 \end{bmatrix} \leftrightarrow \begin{bmatrix} 1 \\ -1 \\ 1 \\ -1 \end{bmatrix} = V_3(k) \qquad \alpha_4 = \begin{bmatrix} 0 \\ 0 \\ 0 \\ 1 \end{bmatrix} \leftrightarrow \begin{bmatrix} 1 \\ j \\ -1 \\ -j \end{bmatrix} = V_4(k)$$

Step 3: Find $[V_i(k)]_\beta$, the coordinates of $V_i(k)$ with respect to the codomain basis β. Since this basis has the usual (nice) orthonormal vectors, the coordinates of each vector are simply the values of the vector itself. To see this, take $V_2(k)$,

$$V_2(k) = a_1\beta_1 + a_2\beta_2 + a_3\beta_3 + a_4\beta_4$$

or

$$\begin{bmatrix} 1 \\ -j \\ -1 \\ j \end{bmatrix} = a_1 \begin{bmatrix} 1 \\ 0 \\ 0 \\ 0 \end{bmatrix} + a_2 \begin{bmatrix} 0 \\ 1 \\ 0 \\ 0 \end{bmatrix} + a_3 \begin{bmatrix} 0 \\ 0 \\ 1 \\ 0 \end{bmatrix} + a_4 \begin{bmatrix} 0 \\ 0 \\ 0 \\ 1 \end{bmatrix}$$

This gives $a_1 = 1$, $a_2 = -j$, $a_3 = -1$, and $a_4 = j$, and these are the values of the vector $V_2(k)$. Therefore, the coordinates $[V_i(k)]_\beta$ are simply the values of $V_i(k)$.

Step 4: The matrix of transformation is given by the columns

$$W = \left[[V_1(k)]_\beta \quad [V_2(k)]_\beta \quad [V_3(k)]_\beta \quad [V_4(k)]_\beta \right] = \begin{bmatrix} 1 & 1 & 1 & 1 \\ 1 & -j & -1 & j \\ 1 & -1 & 1 & -1 \\ 1 & j & -1 & -j \end{bmatrix}$$

This matrix is usually written in a standard form. (See Eqs. 1.10 and 1.11.)

$$W_N = e^{-j2\pi/N} \tag{7.3}$$

The matrix of transformation can be written in terms of W_N as

$$W = \begin{bmatrix} W_4^0 & W_4^0 & W_4^0 & W_4^0 \\ W_4^0 & W_4^1 & W_4^2 & W_4^3 \\ W_4^0 & W_4^2 & W_4^4 & W_4^6 \\ W_4^0 & W_4^3 & W_4^6 & W_4^9 \end{bmatrix} \tag{7.4}$$

The procedure above amounts to the following MATLAB command.

```
w = fft(eye(4))
```
(7.5)

Each column of the identity matrix is a basis vector for the domain. Step 2 finds the DTFS of each column, and since the basis vectors for the codomain are nice orthonormal vectors, the coordinates of $[V_i(k)]_\beta$ are just the components in the transform of each column. The simple MATLAB command in Eq. 7.5 finds the matrix of transformation. This same observation holds true for any N, not just for $N = 4$.

Example 7.1. (a) Find the matrix of transformation for the DTFS applied to a signal of length 3.
(b) Find the transform of $v = \{1 \ -2 \ 1\}$.

Solution: (a) *Step 1*: Choose as bases

$$\alpha = \beta = \left\{ \begin{bmatrix} 1 \\ 0 \\ 0 \end{bmatrix} \begin{bmatrix} 0 \\ 1 \\ 0 \end{bmatrix} \begin{bmatrix} 0 \\ 0 \\ 1 \end{bmatrix} \right\}$$

Step 2: Transform each basis vector by Eq. 7.1.

$$\begin{aligned}
\alpha_1 &= \{1 \ \ 0 \ \ 0\} \leftrightarrow \{1 \ \ 1 \ \ 1\} = V_1(k) \\
\alpha_2 &= \{0 \ \ 1 \ \ 0\} \leftrightarrow \left\{1 \quad -\tfrac{1}{2} - j\tfrac{\sqrt{3}}{2} \quad -\tfrac{1}{2} + j\tfrac{\sqrt{3}}{2}\right\} = V_2(k) \\
\alpha_3 &= \{0 \ \ 0 \ \ 1\} \leftrightarrow \left\{1 \quad -\tfrac{1}{2} + j\tfrac{\sqrt{3}}{2} \quad -\tfrac{1}{2} - j\tfrac{\sqrt{3}}{2}\right\} = V_3(k)
\end{aligned}$$

Steps 3 and 4: The columns of the matrix of transformation consist of $V_1(k)$, $V_2(k)$, and $V_3(k)$.

$$W = \begin{bmatrix} 1 & 1 & 1 \\ 1 & -\tfrac{1}{2} - j\tfrac{\sqrt{3}}{2} & -\tfrac{1}{2} + j\tfrac{\sqrt{3}}{2} \\ 1 & -\tfrac{1}{2} + j\tfrac{\sqrt{3}}{2} & -\tfrac{1}{2} - j\tfrac{\sqrt{3}}{2} \end{bmatrix} = \begin{bmatrix} W_3^0 & W_3^0 & W_3^0 \\ W_3^0 & W_3^1 & W_3^2 \\ W_3^0 & W_3^2 & W_3^4 \end{bmatrix}$$

(b) $$V = Wv = \begin{bmatrix} 1 & 1 & 1 \\ 1 & -\frac{1}{2} - j\frac{\sqrt{3}}{2} & -\frac{1}{2} + j\frac{\sqrt{3}}{2} \\ 1 & -\frac{1}{2} + j\frac{\sqrt{3}}{2} & -\frac{1}{2} - j\frac{\sqrt{3}}{2} \end{bmatrix} \begin{bmatrix} 1 \\ -2 \\ 1 \end{bmatrix} = \begin{bmatrix} 0 \\ \frac{3(1+j\sqrt{3})}{2} \\ \frac{3(1-j\sqrt{3})}{2} \end{bmatrix}$$

This gives two ways to calculate the DTFS. Either apply Eq. 7.1, or find the matrix of transformation W and multiply to get $V = Wv$.

■

Drill 7.1. Find the matrix of transformation for the DTFS applied to a time signal of length 5.

Answer: $W =$

$$\begin{bmatrix} 1 & 1 & 1 & 1 & 1 \\ 1 & 0.31 - j0.95 & -0.81 - j0.59 & -0.81 + j0.59 & 0.31 + j0.95 \\ 1 & -0.81 - j0.59 & 0.31 + j0.95 & 0.31 - j0.95 & -0.81 + j0.59 \\ 1 & -0.81 + j0.59 & 0.31 - j0.95 & 0.31 + j0.95 & -0.81 - j0.59 \\ 1 & 0.31 + j0.95 & -0.81 + j0.59 & -0.81 - j0.59 & 0.31 - j0.95 \end{bmatrix}$$

7.2. Matrix Decomposition View

Here is an example to explain how the FFT algorithm works. This is not a derivation; it simply shows how the FFT works for $N = 4$. A wonderful little book by E. Oran Brigham, *The Fast Fourier Transform* (Prentice-Hall, 1974) gives the details. The first step is to rewrite Eq. 7.4 as

$$W = \begin{bmatrix} 1 & 1 & 1 & 1 \\ 1 & W_4^1 & W_4^2 & W_4^3 \\ 1 & W_4^2 & W_4^0 & W_4^2 \\ 1 & W_4^3 & W_4^2 & W_4^1 \end{bmatrix} \tag{7.6}$$

To derive Eq. 7.6 from 7.4, use the relationship $W_N^{nk} = W_N^{nk \bmod (N)}$. Thus, $W_N^0 = 1$, $W_4^4 = W_4^0 = 1$, $W_4^6 = W_4^2$, and $W_4^9 = W_4^5 = W_4^1$. The next step is to interchange rows 2 and 3 in W and write this matrix as the product of two matrices.

$$\begin{bmatrix} 1 & 1 & 1 & 1 \\ 1 & W_4^2 & W_4^0 & W_4^2 \\ 1 & W_4^1 & W_4^2 & W_4^3 \\ 1 & W_4^3 & W_4^2 & W_4^1 \end{bmatrix} = \begin{bmatrix} 1 & W_4^0 & 0 & 0 \\ 1 & W_4^2 & 0 & 0 \\ 0 & 0 & 1 & W_4^1 \\ 0 & 0 & 1 & W_4^3 \end{bmatrix} \begin{bmatrix} 1 & 0 & W_4^0 & 0 \\ 0 & 1 & 0 & W_4^0 \\ 1 & 0 & W_4^2 & 0 \\ 0 & 1 & 0 & W_4^2 \end{bmatrix} \tag{7.7}$$

You should verify that the product of the two matrices on the right gives the scrambled W matrix on the left. (This scrambling, or interchanging of rows in the W matrix, is part of the FFT algorithm.)

We now wish to demonstrate that using the two matrices on the right reduces the number of operations, as opposed to using the one matrix on the left in Eq. 7.7. Use the matrix in Eq. 7.4 to write the DTFS as

$$\begin{bmatrix} V(0) \\ V(1) \\ V(2) \\ V(3) \end{bmatrix} = \begin{bmatrix} 1 & 1 & 1 & 1 \\ 1 & W_4^1 & W_4^2 & W_4^3 \\ 1 & W_4^2 & W_4^4 & W_4^6 \\ 1 & W_4^3 & W_4^6 & W_4^9 \end{bmatrix} \begin{bmatrix} v(0) \\ v(1) \\ v(2) \\ v(3) \end{bmatrix} \tag{7.8}$$

This requires nine complex multiplications and 12 complex additions. In contrast, consider using the two matrices on the right of Eq. 7.7. First, multiply the v vector by the rightmost matrix to obtain

$$\begin{bmatrix} t(0) \\ t(1) \\ t(2) \\ t(3) \end{bmatrix} = \begin{bmatrix} 1 & 0 & W_4^0 & 0 \\ 0 & 1 & 0 & W_4^0 \\ 1 & 0 & W_4^2 & 0 \\ 0 & 1 & 0 & W_4^2 \end{bmatrix} \begin{bmatrix} v(0) \\ v(1) \\ v(2) \\ v(3) \end{bmatrix} \tag{7.9}$$

Then multiply the resulting t vector by the other matrix to obtain

$$\begin{bmatrix} V(0) \\ V(2) \\ V(1) \\ V(3) \end{bmatrix} = \begin{bmatrix} 1 & W_4^0 & 0 & 0 \\ 1 & W_4^2 & 0 & 0 \\ 0 & 0 & 1 & W_4^1 \\ 0 & 0 & 1 & W_4^3 \end{bmatrix} \begin{bmatrix} t(0) \\ t(1) \\ t(2) \\ t(3) \end{bmatrix} \tag{7.10}$$

Notice that rows 2 and 3 are interchanged in the transform on the left. To get the correct result, we must reshuffle the output terms. Now consider the number of products and sums in these two equations. The first row of Eq. 7.9 gives

$$v(0) + W_4^0 v(2)$$

which is just one complex multiplication and one complex addition. There are eight rows in Eqs. 7.9 and 7.10, giving a total of eight complex multiplications and additions. Thus, we save one multiplication and four additions. More savings occur as N increases. For the DTFS of Eq. 7.8, the number of operations is approximately N^2. For the FFT the number of operations is proportional to $N \log N$.

Example 7.2. Find the DTFS of the four-point sequence $v(n) = \{2, -1, 0, 1\}$ using (a) Eq. 7.6, and (b) 7.7.

Solution: (a) Equation 7.6 gives

$$\begin{bmatrix} V(0) \\ V(1) \\ V(2) \\ V(3) \end{bmatrix} = \begin{bmatrix} 1 & 1 & 1 & 1 \\ 1 & -j & -1 & j \\ 1 & -1 & 1 & -1 \\ 1 & j & -1 & -j \end{bmatrix} \begin{bmatrix} 2 \\ -1 \\ 0 \\ 1 \end{bmatrix} = \begin{bmatrix} 2 \\ 2+j2 \\ 2 \\ 2-j2 \end{bmatrix}$$

(b) Equation 7.9 gives

$$\begin{bmatrix} t(0) \\ t(1) \\ t(2) \\ t(3) \end{bmatrix} = \begin{bmatrix} 1 & 0 & 1 & 0 \\ 0 & 1 & 0 & 1 \\ 1 & 0 & -1 & 0 \\ 0 & 1 & 0 & -1 \end{bmatrix} \begin{bmatrix} 2 \\ -1 \\ 0 \\ 1 \end{bmatrix} = \begin{bmatrix} 2 \\ 0 \\ 2 \\ -2 \end{bmatrix}$$

Plugging this result into Eq. 7.10 gives

$$\begin{bmatrix} V(0) \\ V(2) \\ V(1) \\ V(3) \end{bmatrix} = \begin{bmatrix} 1 & 1 & 0 & 0 \\ 1 & -1 & 0 & 0 \\ 0 & 0 & 1 & -j \\ 0 & 0 & 1 & j \end{bmatrix} \begin{bmatrix} 2 \\ 0 \\ 2 \\ -2 \end{bmatrix} = \begin{bmatrix} 2 \\ 2 \\ 2+j2 \\ 2-j2 \end{bmatrix}$$

Put this answer in proper order to get

$$\begin{bmatrix} V(0) \\ V(1) \\ V(2) \\ V(3) \end{bmatrix} = \begin{bmatrix} 2 \\ 2+j2 \\ 2 \\ 2-j2 \end{bmatrix}$$

■

7.3. Signal Flow Graph Representation

The DTFS is given by

$$V(k) = \sum_{n=0}^{N-1} v(n) e^{-j2\pi nk/N} = \sum_{n=0}^{N-1} v(n) W_N^{nk} \tag{7.11}$$

where

$$W_N = e^{-j2\pi/N}$$

To explain the signal flow graph representation, we begin with the simplest case, $N = 2$.

$N = 2$: Apply Eq. 7.11 to signal $v = \{v(0)\ v(1)\}$ to obtain

$$V(k) = v(0) + W_2^k v(1)$$

or

$$V(0) = v(0) + v(1)$$
$$V(1) = v(0) - v(1)$$

Figure 7.1a depicts the first equation and Fig. 7.1b shows the second equation. These two ways to represent the operations are combined in one graph to obtain the "butterfly" in Fig. 7.1c. Notice that this butterfly represents the matrix equation

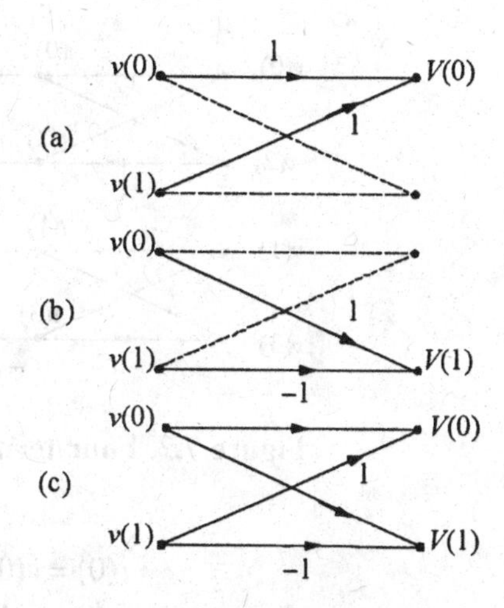

Figure 7.1. DTFS butterfly.

$$\begin{bmatrix} V(0) \\ V(1) \end{bmatrix} = \begin{bmatrix} 1 & W_2^0 \\ 1 & W_2^1 \end{bmatrix} \begin{bmatrix} v(0) \\ v(1) \end{bmatrix}$$

$N = 4$

Apply Eq. 7.11 to obtain

$$V(k) = v(0) + W_4^k v(1) + W_4^{2k} v(2) + W_4^{3k} v(3) \qquad (7.12)$$

To break this equation into the two parts corresponding to Eqs. 7.9 and 7.10, note that $W_4^{2k} = W_2^k$. This allows Eq. 7.12 to be written as

$$V(k) = v(0) + v(2)W_2^k + W_4^k [v(1) + v(3)W_2^k] \qquad (7.13)$$

This gives the following four equations, one for each k.

$$\begin{aligned} V(0) &= v(0) + v(2) + v(1) + v(3) \\ V(1) &= v(0) - v(2) - j[v(1) - v(3)] \\ V(2) &= v(0) + v(2) - [v(1) + v(3)] \\ V(3) &= v(0) - v(2) + j[v(1) - v(3)] \end{aligned} \qquad (7.14)$$

Figure 7.2 shows the signal flow graph for these four equations. The left side of the graph represents Eq. 7.9, and the right side represents Eq. 7.10. The intermediate variables t are given by

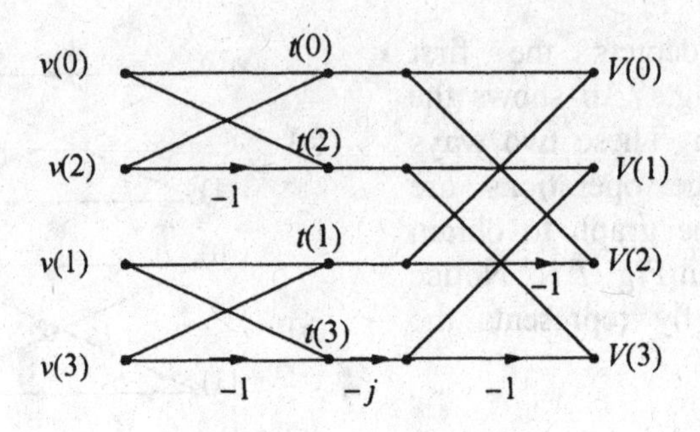

Figure 7.2. Four-term signal flow graph.

$$t(0) = v(0) + v(2)$$
$$t(1) = v(1) + v(3)$$
$$t(2) = v(0) - v(2)$$
$$t(3) = v(1) - v(3)$$

In matrix form these equations become

$$\begin{bmatrix} t(0) \\ t(1) \\ t(2) \\ t(3) \end{bmatrix} = \begin{bmatrix} 1 & 0 & 1 & 0 \\ 0 & 1 & 0 & 1 \\ 1 & 0 & -1 & 0 \\ 0 & 1 & 0 & -1 \end{bmatrix} \begin{bmatrix} v(0) \\ v(1) \\ v(2) \\ v(3) \end{bmatrix}$$

This agrees with Eq. 7.9. Plugging these variables into Eqs. 7.14 gives the right side of Fig. 7.2.

$$V(0) = t(0) + t(1)$$
$$V(1) = t(2) - j\,t(3)$$
$$V(2) = t(0) - t(1)$$
$$V(3) = t(2) + j\,t(3)$$

In matrix form, these equations agree with Eq. 7.10.

$$\begin{bmatrix} V(0) \\ V(2) \\ V(1) \\ V(3) \end{bmatrix} = \begin{bmatrix} 1 & 1 & 0 & 0 \\ 1 & -1 & 0 & 0 \\ 0 & 0 & 1 & -j \\ 0 & 0 & 1 & j \end{bmatrix} \begin{bmatrix} t(0) \\ t(1) \\ t(2) \\ t(3) \end{bmatrix}$$

The Basic Idea

The basic idea behind the FFT is to break the DTFS into parts, work on each part separately, and then combine the results. Of course, one must "dissect the bird at its joints" (i.e., select the parts so the results can easily be combined). An explanation of how to select the parts begins with a particular data storage scheme.

Consider the computation of an N-point DTFS, where N is a composite number given by

$$N = LM \tag{7.15}$$

(A composite number can be represented by the product of two numbers other than itself and 1.) The sequence $v(n)$, $n = 0, 1, \ldots, N-1$, can be stored in a one-dimensional array of length N, or it can be stored in a two-dimensional array of size $L \times M$. Figure 7.3 shows the relation between $v(n)$ and $x(\ell,m)$, where

$$x(0,0) = v(0),\ x(0,1) = v(1),\ \ldots,\ x(L-1, M-1) = v(N-1)$$

This represents a row-wise storage scheme, where the indices are related by

$$n = M\ell + m \tag{7.16}$$

Note that the data could also be stored column-wise with the relation

$$n = Lm + \ell \tag{7.17}$$

The following development stores the input data $\{v(n)\}$ column-wise according to Eq. 7.17, and stores the output data $\{V(k)\}$ row-wise. Use the frequency-domain indices k and (p,q), where

$$k = Mp + q \tag{7.18}$$

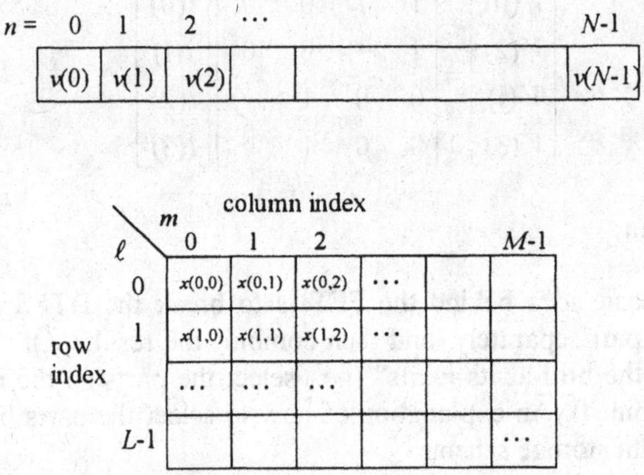

$n =$	0	1	2	...		N-1
	$v(0)$	$v(1)$	$v(2)$			$v(N$-1$)$

column index

ℓ \ m	0	1	2			M-1
0	$x(0,0)$	$x(0,1)$	$x(0,2)$	...		
1	$x(1,0)$	$x(1,1)$	$x(1,2)$	...		
row index	...	...	...			
L-1						...

Figure 7.3. Storing $v(n)$ in a two-dimensional array.

This gives the relationship

$$V(0) = X(0,0),\ V(1) = X(0,1),\ \ldots,\ V(N-1) = X(L-1, M-1)$$

With the column-wise mapping for $v(n)$ and the row-wise mapping for $V(k)$, the DTFS formula in terms of $x(\ell,m)$ and $X(p,q)$ becomes

$$X(p,q) = \sum_{m=0}^{M-1} \sum_{\ell=0}^{L-1} x(\ell,m) W_N^{(Mp+q)(Lm+\ell)} \tag{7.19}$$

Simplify this by writing

$$W_N^{(Mp+q)(Lm+\ell)} = W_N^{MLmp} W_N^{Mp\ell} W_N^{Lmq} W_N^{\ell q}$$

But

$$W_N^{MLmp} = W_N^{Nmp} = 1,$$
$$W_N^{Mp\ell} = W_{N/M}^{p\ell} = W_L^{p\ell},$$
$$W_N^{Lmq} = W_{N/L}^{mq} = W_M^{mq}$$

Substitute these simplifications into Eq. 7.19 to get

$$X(p,q) = \sum_{\ell=0}^{L-1} \left\{ W_N^{\ell q} \left[\sum_{m=0}^{M-1} x(\ell,m) W_M^{mq} \right] \right\} W_L^{p\ell} \tag{7.20}$$

To understand this expression, break the computation into three steps:

Step 1: Evaluate the expression

$$T(\ell,q) = \sum_{m=0}^{M-1} x(\ell,m) W_M^{mq}, \quad q = 0, 1, \cdots, M-1 \tag{7.21}$$

for each row. There are M values of the DTFS for each row, and there are L rows.

Step 2: Compute a new rectangular array $R(\ell,q)$ defined by

$$R(\ell,q) = W_N^{\ell q} T(\ell,q), \quad 0 \le \ell \le L-1, \quad 0 \le q \le M-1 \tag{7.22}$$

Step 3: Complete the calculation by evaluating

$$X(p,q) = \sum_{\ell=0}^{L-1} R(\ell,q) W_L^{p\ell} \tag{7.23}$$

Let us examine each step in detail. Suppose that we wish to find the DTFS of a 15-point sequence $v(n)$, where

$$v(n) = \{ v(0), v(1), \ldots, v(14) \}$$

Since 15 factors into 3×5, store $v(n)$ column-wise in a 3×5 array.

$$x = \begin{bmatrix} v(0) & v(5) & v(10) \\ v(1) & v(6) & v(11) \\ v(2) & v(7) & v(12) \\ v(3) & v(8) & v(13) \\ v(4) & v(9) & v(14) \end{bmatrix}$$

Step 1: Calculate the 3-point DTFS of each row to find the T-matrix.

$$T = \begin{bmatrix} t(0,0) & t(0,1) & t(0,2) \\ t(1,0) & t(1,1) & t(1,2) \\ t(2,0) & t(2,1) & t(2,2) \\ t(3,0) & t(3,1) & t(3,2) \\ t(4,0) & t(4,1) & t(4,2) \end{bmatrix}$$

This means that the T matrix is derived by taking the transform of each row of the x matrix, that is,

$$\begin{aligned} &\{v(0) \quad v(5) \quad v(10)\} \leftrightarrow \{t(0,0) \quad t(0,1) \quad t(0,2)\} \\ &\{v(1) \quad v(6) \quad v(11)\} \leftrightarrow \{t(1,0) \quad t(1,1) \quad t(1,2)\} \\ &\qquad \cdots \qquad\qquad\quad \cdots \qquad\qquad \cdots \\ &\{v(4) \quad v(9) \quad v(14)\} \leftrightarrow \{t(4,0) \quad t(4,1) \quad t(4,2)\} \end{aligned}$$

Step 2: Multiply the T matrix term by term by the upper left corner of W_N, which we call W_U, to produce the R matrix. The relation between W_N and W_U is pictured in Fig. 7.4.

Figure 7.5 shows this term-by-term product operation. In other words, this is not matrix multiplication. This operation produces the R matrix, given by

$$R = \begin{bmatrix} r(0,0) & r(0,1) & r(0,2) \\ r(1,0) & r(1,1) & r(1,2) \\ r(2,0) & r(2,1) & r(2,2) \\ r(3,0) & r(3,1) & r(3,2) \\ r(4,0) & r(4,1) & r(4,2) \end{bmatrix}$$

Step 3: Find the five-point DTFS of each column in R to produce the output X matrix. This means that

$$\begin{aligned} &\{r(0,0) \quad r(1,0) \quad r(2,0) \quad r(3,0) \quad r(4,0)\} \leftrightarrow \\ &\qquad \{X(0,0) \quad X(1,0) \quad X(2,0) \quad X(3,0) \quad X(4,0)\} \end{aligned}$$

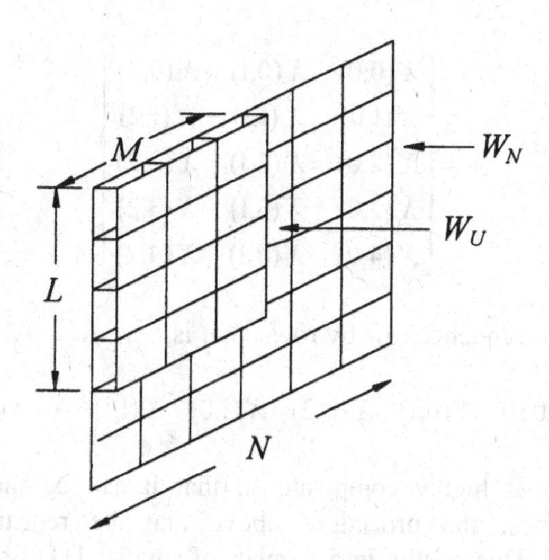

Figure 7.4. W_N and W_U.

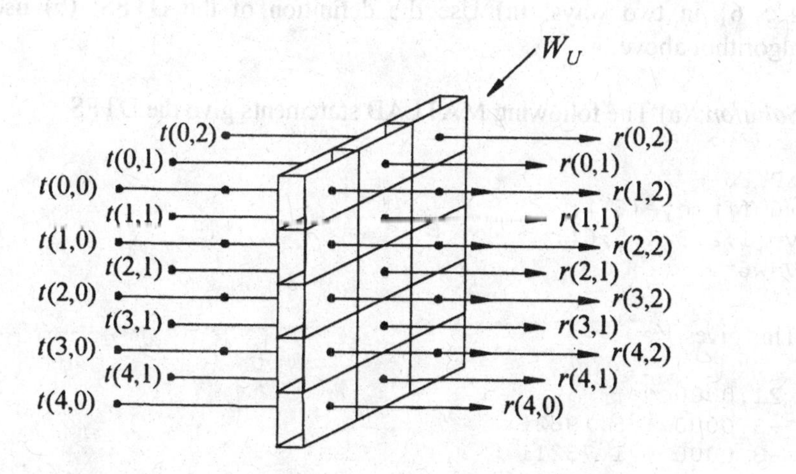

Figure 7.5. Term-by-term multiplication operation.

The same scheme applies to the other two columns of the R matrix. The output X matrix contains the transform of the input sequence $v(n)$ in somewhat jumbled form. The X matrix has the form

$$X = \begin{bmatrix} X(0,0) & X(0,1) & X(0,2) \\ X(1,0) & X(1,1) & X(1,2) \\ X(2,0) & X(2,1) & X(2,2) \\ X(3,0) & X(3,1) & X(3,2) \\ X(4,0) & X(4,1) & X(4,2) \end{bmatrix}$$

Read the output V sequence row by row. That is,

$$V = \{X(0,0) \quad X(0,1) \quad X(0,2) \quad X(1,0) \quad X(1,1) \quad \cdots \quad x(4,2)\}$$

When N is highly composite so that it can be factored into a number of primes, the procedure above may be repeated for each additional prime. This results in a number of smaller DTFSs, which leads to computational savings.

Example 7.3. Use MATLAB to calculate the DTFS of the array $v = \{1\ 2\ 3\ 4\ 5\ 6\}$ in two ways: (a) Use the definition of the DTFS; (b) use the algorithm above.

Solution: (a) The following MATLAB statements give the DTFS.

```
%DTFS
w6=fft(eye(6));
v=[1;2;3;4;5;6];
V=w6*v
```

This gives V =

```
 21.0000
 -3.0000 + 5.1962i
 -3.0000 + 1.7321i
 -3.0000 - 0.0000i
 -3.0000 - 1.7321i
 -3.0000 - 5.1962i
```

(b) Now let's see if the FFT algorithm gives the same answer. First store the data column-wise in a 3×2 array, perform Steps 1, 2, and 3 above, then read the output data row-wise. The following statements store array v column-wise and define some necessary variables and matrices.

```
L=3;
M=2;
x=[v(1:3) v(4:6)];
w3=fft(eye(3));
w2=fft(eye(2));
```

Step 1: Evaluate T by the statement

```
t=x*w2;
```

Step 2: Multiply T by the upper left corner of W_6.

```
w6=fft(eye(6));
u=w6(1:L,1:M);
r=t.*u;
```

Step 3: Calculate X.

```
X=w3*r
```

The output is `X =`

```
 21.0000        -3.0000 + 5.1962i
 -3.0000 + 1.7321i -3.0000 - 0.0000i
 -3.0000 - 1.7321i -3.0000 - 5.1962i
```

Read the output row-wise. This agrees with the answer to part (a).

■

Steps 1 through 3 can be used to draw the signal flow graph. Let us show how to do this for a six-point signal in order to further illustrate the FFT.

Example 7.4. Draw the signal flow graph for a six-point FFT.

Solution: Store the input data column-wise in a 3×2 matrix like this:

$$x = \begin{bmatrix} v(0) & v(3) \\ v(1) & v(4) \\ v(2) & v(5) \end{bmatrix}$$

Notice that the data are stored column-wise. Next, find the DTFS of each row to obtain the T matrix. Then multiply the T matrix point by point by W_U, the upper left portion of W_6. The matrix W_U is given by

$$W_U = \begin{bmatrix} W_6^0 & W_6^0 \\ W_6^0 & W_6^1 \\ W_6^0 & W_6^2 \end{bmatrix} = \begin{bmatrix} 1 & 1 \\ 1 & e^{-j\pi/3} \\ 1 & e^{-j2\pi/3} \end{bmatrix}$$

This gives the graph in Fig. 7.6.

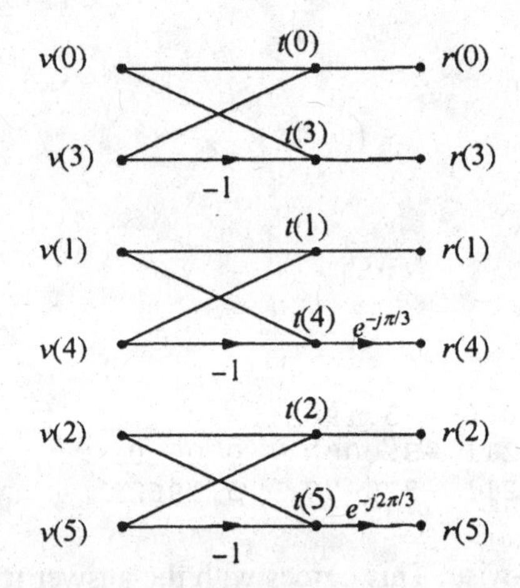

Figure 7.6. Steps 1 and 2.

The next step is to construct two three-point DTFSs of each column of the R matrix. Figure 7.7 shows the first-column DTFS.

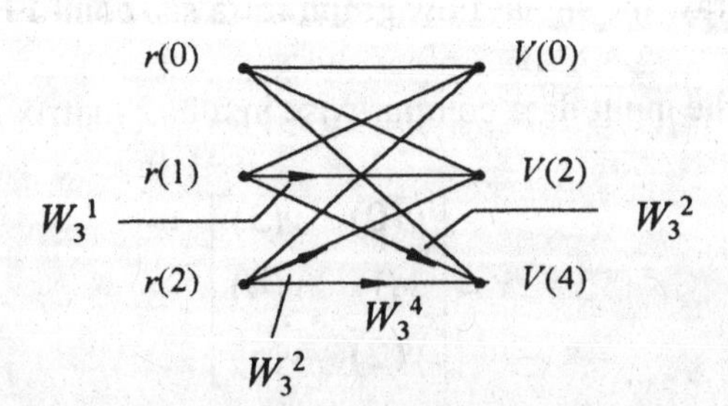

Figure 7.7. Three-point DTFS applied to the first column.

The second-column DTFS is identical except that the inputs are $r(3)$, $r(4)$, and $r(5)$. The outputs are $V(1)$, $V(3)$, and $V(5)$, respectively. The complete signal flow graph in Fig. 7.8 is the result of combining Fig. 7.6 with the two versions of the three-point DTFS. The first version is connected to the points labeled $r(0)$, $r(1)$, and $r(2)$, and the second version is connected to $r(3)$, $r(4)$, and $r(5)$ in Fig. 7.6.

■

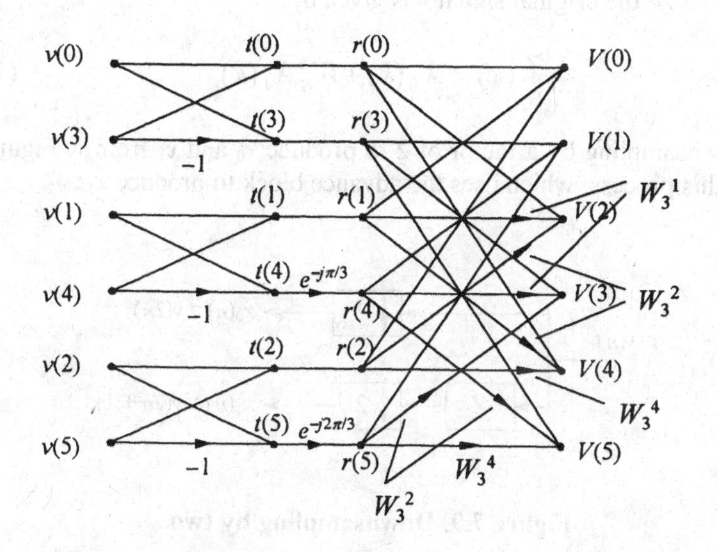

Figure 7.8. Signal flow graph for the six-point FFT.

7.4. Downsampling View

For a signal v of length N, where N is an integer power of 2, break this sequence into two parts:

$$x_0 = \{v(0) \quad v(2) \quad v(4) \quad \cdots \quad v(N-2)\}$$
$$x_1 = \{v(1) \quad v(3) \quad v(5) \quad \cdots \quad v(N-1)\}$$

The sequences x_0 and x_1 have length $N/2$, so the DTFS of each sequence is given by

$$X_0(k) = \sum_{n=0}^{\frac{N}{2}-1} x_0(n) W_{N/2}^{nk} = \sum_{n=0}^{\frac{N}{2}-1} v(2n) W_{N/2}^{nk} \tag{7.24}$$

$$X_1(k) = \sum_{n=0}^{\frac{N}{2}-1} x_1(n) W_{N/2}^{nk} = \sum_{n=0}^{\frac{N}{2}-1} v(2n+1) W_{N/2}^{nk} \tag{7.25}$$

The DTFS of the original signal v is given by

$$V(k) = X_0(k) + W_N^k X_1(k) \tag{7.26}$$

Use downsampling by a factor of 2 to produce x_0 and x_1 from v. Figure 7.9 shows this process, which uses the advance block to produce x_1.

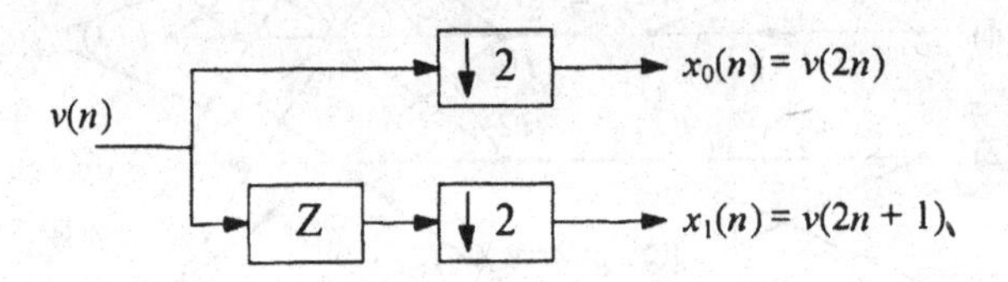

Figure 7.9. Downsampling by two.

This downsampling scheme provides the ability to produce the sequences x_0 and x_1, allowing the DTFS of Eq. 7.26 to be constructed. Figure 7.10 shows this, where the two DTFS blocks represent X_0 and X_1, respectively. That is, the operations in Eqs. 7.24 and 7.25 are represented by the two $N/2$ DTFS blocks.

The next step replaces each $N/2$ DTFS block by the diagram in Fig. 7.10. This process continues until we reach the limit, $N = 2$. This gives a nested set of DTFS operations, much like the sequence of images that one obtains by holding two mirrors so that they face each other. This nested set of DTFS operations produces the FFT.

This downsampling view is just a different way of looking at the FFT. The scheme depicted in Fig. 7.10 is called the *decimation-in-time algorithm*. Here is an example to illustrate this view.

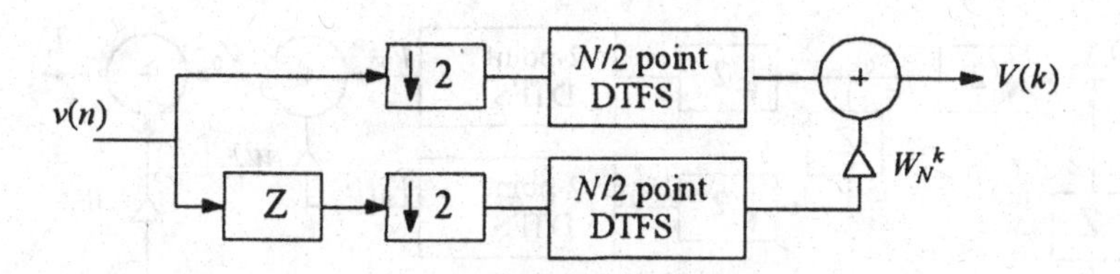

Figure 7.10. DTFS operation.

Example 7.5. Use MATLAB to calculate the DTFS of the array

$$v = \{1, 2, 3, 4, 5, 6, 7, 8\}$$

(a) Use the definition of the DTFS.
(b) Use downsampling.

Solution: (a)

```
w8=fft(eye(8));
v=[1; 2; 3; 4; 5; 6; 7; 8];
V=w8*v
```

This produces the array

$$V = [36 \;\; (-4 + j9.657) \;\; (-4 + j4) \;\; (-4 + j1.657) \;\; -4 \;\; (-4 - j1.657) \;\; (-4 - j4) \;\; (-4 - j9.657)]$$

(b) Figure 7.11 shows the complete scheme for calculating an eight-point FFT. The eight-point input signal is labeled v. The output of the first downsampling stage consists of v_0 and v_1, each of length 4. The signals v_{00}, v_{01}, v_{10}, and v_{11} each have length 2. These are the input to the two-point DTFS operations.

Figure 7.12 shows two views of the two-point FFT operation, first as the butterfly and then as downsampling. Either view is valid, for they amount to the same thing. Now here are the details of the FFT, beginning with the eight-point signal $v(n$.)

Step 1. Downsample $v(n)$ to obtain two four-point signals, $v_0(n)$ and $v_1(n)$.

$$v_0 = \{1, 3, 5, 7\} \qquad v_1 = \{2, 4, 6, 8\}$$

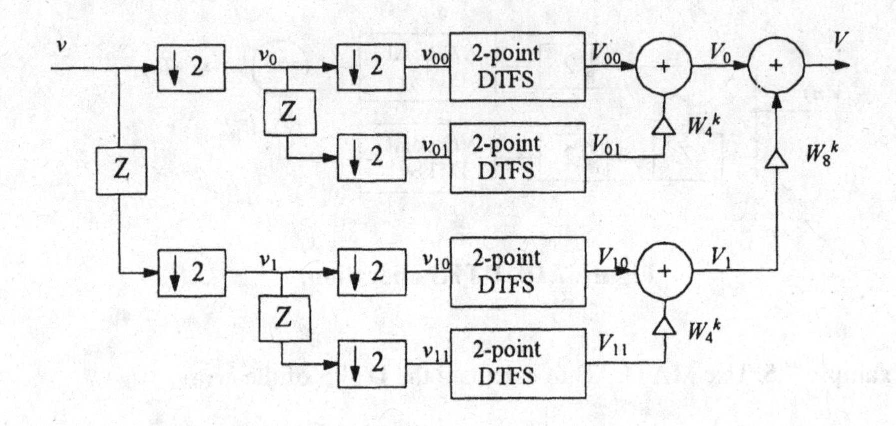

Figure 7.11. Eight-point FFT as a downsampling operation.

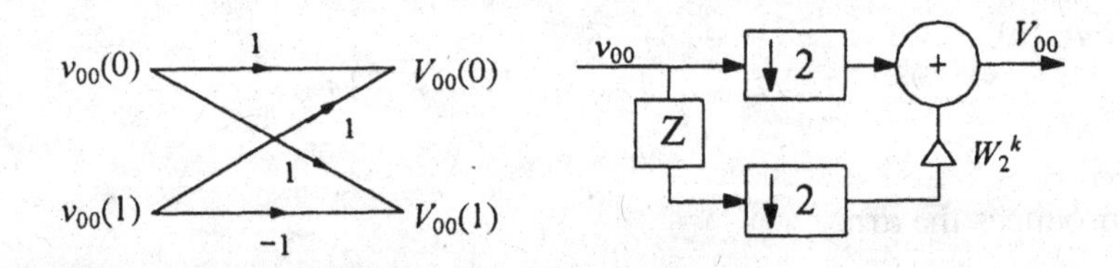

Figure 7.12. Two views of the two-point FFT.

Step 2: Downsample each four-point signal to obtain

$$v_{00} = \{1, 5\} \qquad v_{01} = \{3, 7\} \qquad v_{10} = \{2, 6\} \qquad v_{11} = \{4, 8\}$$

Step 3: Apply the two-point FFT to each signal to obtain

$$V_{00} = \{6, -4\} \qquad V_{01} = \{10, -4\}$$
$$V_{10} = \{8, -4\} \qquad V_{11} = \{12, -4\}$$

Step 4: Combine V_{00} with V_{01} according to Eq. 7.26:

$$V_0(k) = V_{00}(k) + W_4^k V_{01}(k)$$

$$V_0(k) = V_{00}(k) + W_4^k V_{01}(k)$$

Note that this is cyclic, or $V_{00}(2) = V_{00}(0)$ and $V_{00}(3) = V_{00}(1)$. The same statement applies to V_{01}, V_{10}, and V_{11}. The four values of V_0 are given by

$$\begin{aligned}
V_0(0) &= V_{00}(0) + W_4^0 V_{01}(0) = 6 + 10 = 16 \\
V_0(1) &= V_{00}(1) + W_4^1 V_{01}(1) = -4 + j4 \\
V_0(2) &= V_{00}(0) + W_4^2 V_{01}(0) = 6 - 10 = -4 \\
V_0(3) &= V_{00}(1) + W_4^3 V_{00}(1) = -4 - j4
\end{aligned}$$

Step 5: Combine V_{10} with V_{11} according to Eq. 7.26.

$$\begin{aligned}
V_1(0) &= V_{10}(0) + W_4^0 V_{11}(0) = 8 + 12 = 20 \\
V_1(1) &= V_{10}(1) + W_4^1 V_{11}(1) = -4 + j4 \\
V_1(2) &= V_{10}(0) + W_4^2 V_{11}(0) = 8 - 12 = -4 \\
V_1(3) &= V_{10}(1) + W_4^3 V_{10}(1) = -4 - j4
\end{aligned}$$

This gives V_0 and V_1 as

$$\begin{aligned}
V_0 &= \{16 \quad (-4 + j4) \quad -4 \quad (-4 - j4)\} \\
V_1 &= \{20 \quad (-4 + j4) \quad -4 \quad (-4 - j4)\}
\end{aligned}$$

Step 6: Combine V_0 with V_1 according to Eq. 7.26 to obtain V. In this instance, Eq. 7.26 has the form

$$V(k) = V_0(k) + W_8^k V_1(k)$$

Thus

$$\begin{aligned}
V(0) &= V_0(0) + W_8^0 V_1(0) = 16 + 20 = 36 \\
V(1) &= V_0(1) + W_8^1 V_1(1) = (-4 + j4) + 1\angle -45°(-4 + j4) \\
&= -4 + j9.657 \\
V(2) &= V_0(2) + W_8^2 V_1(2) = -4 + j4 \\
V(3) &= V_0(3) + W_8^3 V_1(3) = (-4 - j4) + 1\angle -135°(-4 - j4) \\
&= -4 + j1.657
\end{aligned}$$

$$V(4) = V_0(0) + W_8^4 V_1(0) = 16 - 20 = -4$$
$$V(5) = V_0(1) + W_8^5 V_1(1) = (-4 + j4) + 1\angle 135°(-4 + j4)$$
$$= -4 - j1.657$$
$$V(6) = V_0(2) + W_8^6 V_1(2) = -4 - j4$$
$$V(7) = V_0(3) + W_8^7 V_1(3) = (-4 - j4) + 1\angle 45°(-4 - j4)$$
$$= -4 - j9.657$$

This is the same array as that obtained from part (a) of this example. ■

Chapter 8
Wavelet Transform

The wavelet transform differs from other transforms in two ways. The Laplace transform basis consists of a set of exponentials $\{e^{st}\}$. The z transform basis is also a set of exponentials, $\{z^n\}$. Each form of the Fourier transform has its own basis. If we follow the custom of associating a particular transform with its basis, there are many wavelet transforms.

A second difference occurs in the calculation of the transform. Most transforms use the dot product between a waveform $v(t)$ and the basis functions. This works for wavelet transforms if an analytical expression for the wavelet basis functions is known. But there is no formula for most wavelets, so another method must be used. This other method uses the concepts of Chapter 6, and we will have much more to say about that after introducing the Haar wavelets in this chapter.

Chapter Goals: After completing this chapter, you should be able to do the following:

- For a given vector space V_1 and a subset $V_0 \subset V_1$, find the orthogonal complement $V_0^{\perp}$.
- Approximate a given signal to a particular resolution using the Haar wavelet system.
- Use the two-scale relations (Eqs. 8.13 and 8.14) to calculate the lower-resolution scaling function and wavelets.

8.1. Scaling Functions and Wavelets

Begin with a review. First and foremost, a vector can be other things besides a directed magnitude or an $n \times 1$ matrix. A real mathematician thinks of functions (waveforms) when they hear the word *vector*. Recall the definition from Chapter 2, repeated here.

Definition 2.1. A *vector space* is a set $V = \{v_i\}$ together with a field of scalars $A = \{a_i\}$ that has the following two operations and seven properties.

1. Two vectors can add together to obtain a third vector. This mechanism combines two vectors to produce a third.
2. A scalar can multiply a vector to obtain another vector. This mechanism combines a scalar with a vector to produce another vector.

Using these two operations, vector addition and scalar multiplication, the following properties must hold for all $v_i \in V$ and all $a_i \in A$.

(1) $v_1 + v_2 = v_2 + v_1$.
(2) $(v_1 + v_2) + v_3 = v_1 + (v_2 + v_3)$.
(3) $a_1(v_1 + v_2) = a_1v_1 + a_1v_2$.
(4) $(a_1 + a_2)v_i = a_1v_i + a_2v_i$.
(5) $(a_1\, a_2)v_i = a_1\,(a_2v_i)$.
(6) $1 \cdot v_i = v_i$.
(7) There exists a unique vector v_0, called the *zero vector*, such that for all vectors v_i we have $0 \cdot v_i = v_0$.

Now think of waveforms. To be specific, consider continuous-time energy signals. (Recall them as typically being finite in duration.) Any two of them will add together to get a third energy signal. Multiplying by scalars (either real or complex numbers) changes the amplitude of a signal. And energy signals satisfy all seven properties, so energy signals satisfy the requirements to be called a vector space. This means that each energy signal is a vector. Furthermore, an energy signal is just as good a vector as a directed magnitude or a component vector.

In a similar manner, the set of all continuous-time power signals forms another vector space, as do the set of all discrete-time energy signals, or the set of all discrete-time power signals. And there are many other vector spaces. Here are some examples.

Example 8.1. Figure 8.1 shows a piecewise-constant waveform defined over the interval [0, 1]. There are four constant segments, [0, ¼], [¼, ½], [½, ¾], and [¾, 1]. Let V be the set of all such waveforms (i.e., of duration 1 with four constant segments). Define the inner product as

$$\langle v_1 | v_2 \rangle = \int_0^1 v_1(t) v_2(t)\, dt$$

(a) Show that V is a vector space.
(b) Find a basis for V.
(c) Determine the dimension of V.

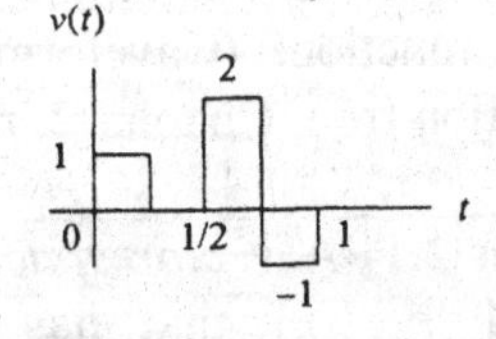

Figure 8.1. Piecewise-constant function (a vector).

Solution: (a) This set satisfies the two operations in Definition 2.1, because we can add together any two piecewise-constant functions and obtain another piecewise-constant function, and we can multiply any function by a scalar and obtain another piecewise-constant function. This set also satisfies the seven properties. For example, the zero vector is zero in each segment.
(b) Figure 8.2 shows one possible basis, and Fig. 8.3 shows another. These two basis sets have several good properties. The basis $\alpha = \{\alpha_1, \alpha_2, \alpha_3, \alpha_4\}$ in Fig. 8.2 is orthogonal. This means that the inner product is zero for any two different basis vectors. However, this basis is not orthonormal, because the inner product of α_i with itself is not 1 for all i. (It would be orthonormal if each waveform had height 2.) The basis in Fig. 8.3 is orthonormal, because the inner product between different vectors is zero, while the inner product of any basis vector with itself is 1.
(c) The dimension of V is 4 because there are four basis vectors.

■

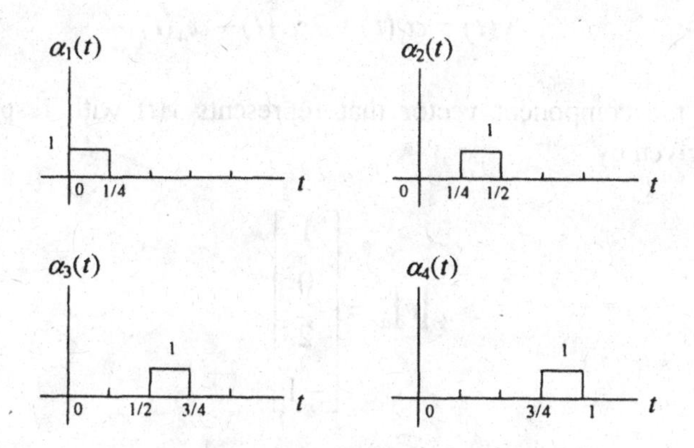

Figure 8.2. One possible basis for V.

Example 8.2. Express $v(t)$ in Fig. 8.1 by its components with respect to (a) the basis α and (b) the basis β.

Solution: (a) Compare the signal in Fig. 8.1 to the four basis vectors in α and see by inspection that $v(t)$ can be expressed as a linear combination of the basis vectors.

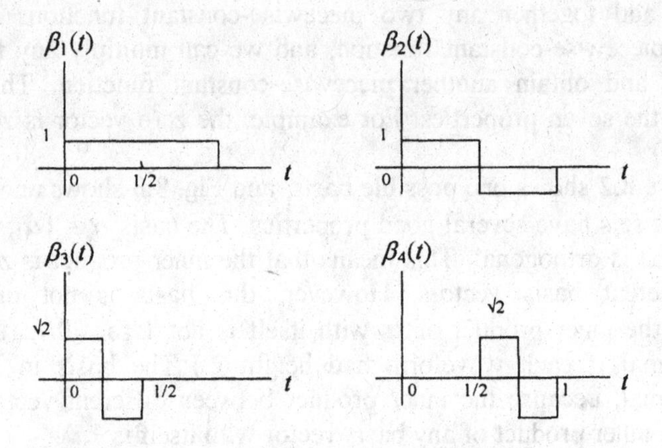

Figure 8.3. Orthonormal basis for *V*.

$$v(t) = \alpha_1(t) + 2\alpha_3(t) - \alpha_4(t)$$

Therefore, the component vector that represents $v(t)$ with respect to the basis α is given by

$$[v]_\alpha = \begin{bmatrix} 1 \\ 0 \\ 2 \\ -1 \end{bmatrix}$$

(b) Here the solution is not so obvious, requiring a systematic approach. The component vector $[v]_\beta$ consists of the four components b_1, b_2, b_3, and b_4, from the relation

$$v(t) = b_1\beta_1(t) + b_2\beta_2(t) + b_3\beta_3(t) + b_4\beta_4(t) \qquad (8.1)$$

which represents $v(t)$ as a linear combination of the basis vectors in β. There are four unknowns, b_1, b_2, b_3, and b_4. To develop four independent equations, take the inner product of each term in this equation with the four basis vectors

$$\langle \beta_i | v \rangle = b_1 \langle \beta_i | \beta_1 \rangle + b_2 \langle \beta_i | \beta_2 \rangle + b_3 \langle \beta_i | \beta_3 \rangle + b_4 \langle \beta_i | \beta_4 \rangle \text{ for } i = 1, 2, 3, 4$$

(The other way, $\langle v | \beta_i \rangle$, would work just as well.) Figure 8.4 shows the waveforms $\beta_1(t)$ and $v(t)$. Multiply the two functions and find the area under their product. This inner product is ¼[1 + 2 − 1] = ½. In a similar manner, the other coefficients give the component vector

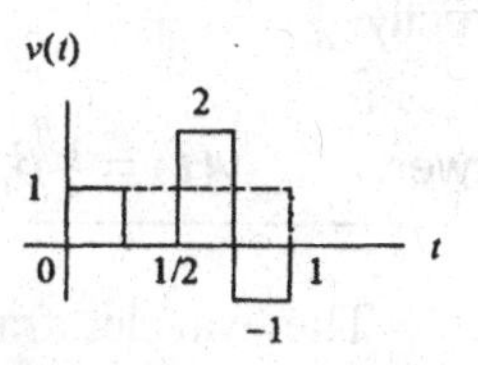

Figure 8.4. The waveforms $\beta_1(t)$ and $v(t)$.

$$[v]_\beta = \begin{bmatrix} \frac{1}{2} \\ 0 \\ \frac{\sqrt{2}}{4} \\ \frac{3\sqrt{2}}{4} \end{bmatrix}$$

Equation 8.1 gives $v(t) = \frac{1}{2}\beta_1(t) + \frac{\sqrt{2}}{4}\beta_3(t) + \frac{3\sqrt{2}}{4}\beta_4(t)$. ■

Here is a new idea. Think of the coefficient vectors as the transform of $v(t)$. Thus, $[v]_\alpha$ is the transform of $v(t)$ with respect to the basis α, and $[v]_\beta$ is the transform of $v(t)$ with respect to the basis β. Write this as

$$v(t) \overset{\alpha}{\leftrightarrow} \{1, 0, 2, -1\} \tag{8.2}$$

$$v(t) \overset{\beta}{\leftrightarrow} \{\tfrac{1}{2}, 0, \tfrac{\sqrt{2}}{4}, \tfrac{3\sqrt{2}}{4}\} \tag{8.3}$$

Equation 8.3 is the Haar wavelet transform of $v(t)$. Haar functions are piecewise-constant waveforms, and we will use these to illustrate the general properties of wavelets before expanding our discussion to include other wavelets.

It is legitimate to call this a transform. Recall our discussion of transforms in Chapter 1. A *transform* (or *operator*) is a relationship between two sets of functions (vectors). In our applications the domain, or input to the black box in Fig. 1.2, consists of time functions. The output vectors in the codomain are called the transform, and in our case these are the wavelet transforms of the input time functions.

Drill 8.1. Find the Haar wavelet transform of $v(t)$ in Fig. 8.5. Use the basis functions in Fig. 8.3. Plot your answer and check that it depicts $v(t)$ correctly.

Answer: $v(t)=\frac{1}{4}\beta_1(t)-\frac{1}{4}\beta_2(t)-\frac{\sqrt{2}}{2}\beta_3(t)+\frac{3\sqrt{2}}{4}\beta_4(t)$

The wavelet transform of a signal $v(t)$ is the set of projections of $v(t)$ onto a basis. (The transform $\{\frac{1}{2}, 0, \frac{\sqrt{2}}{4}, \frac{3\sqrt{2}}{4}\}$ in Eq. 8.3 is the set of projections of $v(t)$ onto the basis vectors β.) This basis is a family of functions that are normalized dilations and translations of a prototype wavelet $\psi(t)$.

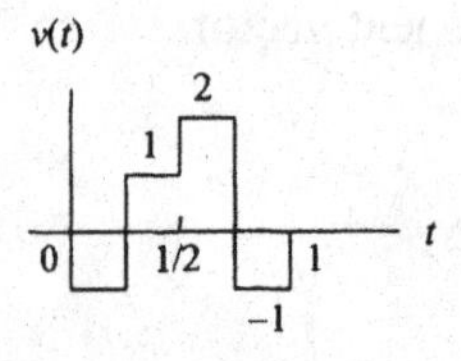

Figure 8.5. Function for Drill 8.1.

(The prototype or mother wavelet in β is β_2.) Furthermore, another set of functions $\varphi(t)$, called the *scaling functions*, is involved. (β_1 in Fig. 8.3 is the prototype scaling function.) And there is a relationship between these two sets of functions that involves the orthogonal complement of a set (defined below). To understand this properly requires knowledge of the direct sum of two vector spaces, so let us begin with some mathematics.

Let V_1 be a vector space, and let V_0 be a subset of V_1, written $V_0 \subset V_1$. Note that V_0 may or may not be a vector space itself. For V_0 to be a subspace (as opposed to a subset), it must satisfy all the properties in the definition for a vector space in Definition 2.1. With this in mind, here is the definition for the orthogonal complement of a set.

Definition 8.1. If V_0 is any subset of a vector space V_1, we define the *orthogonal complement* of V_0, denoted $V_0^\perp$ and pronounced "V_0-perp," as the set of all vectors of V_1 which are orthogonal to each element of V_0; that is,

$$V_0^\perp = \left\{ w \in V_1 \middle| \langle w | v \rangle = 0 \quad \text{for all } v \in V_0 \right.$$

Example 8.3. Let $V_1 = R^2$, the set of all 2×1 matrices. Let $V_0 = \left\{ \begin{bmatrix} 1 \\ 1 \end{bmatrix} \right\}$, a set consisting of a single matrix. Define the dot product by $\langle w|v \rangle = w'v$. (Recall that the inner product may be defined in different ways for the same vector space. The only requirement is that the inner product must satisfy the properties of Definition 2.4.) Find $V_0^{\perp}$ and show that it is a subset of V_1.

Solution: Every vector w that is orthogonal to $v = \begin{bmatrix} 1 \\ 1 \end{bmatrix}$ (including the zero vector) must have the form $w = a\begin{bmatrix} 1 \\ -1 \end{bmatrix}$, meaning that the orthogonal complement is given by $V_0^{\perp} = \left\{ \begin{bmatrix} a \\ -a \end{bmatrix} \right\}$. This set is a vector space because $0_V \in V_0^{\perp}$ and all vectors of the form $a_1 w_1 + a_2 w_2$ are also elements of $V_0^{\perp}$. ■

Example 8.4. Let V_1 be the set of all polynomials of degree 2 or less. Thus

$$x_i(t) = a_i + b_i t + c_i t^2 \quad \text{for all } i$$

are all elements of V_1. Use the inner product $\langle x_1 | x_2 \rangle = a_1 a_2 + b_1 b_2 + c_1 c_2$. (This is a different inner product than in Example 8.3.) Let us arbitrarily choose $V_0 = \{ct^2\}$ for all c. V_0 contains $x_1(t) = 2t^2$ and $x_2(t) = -1.5t^2$, but not $x_3(t) = t + 2t^2$ nor $x_4(t) = 5 - 2t$. Notice that $x_4(t)$ is orthogonal to both $x_1(t)$ and $x_2(t)$, while $x_3(t)$ is not (because of the way the inner product is defined). $V_0^{\perp}$ consists of all polynomials of the form $a + bt$, the set of all polynomials of degree 1 or less. Note that $V_0^{\perp}$ is a subspace of V_1 since it consists of all polynomials of degree 1 or less. ■

It is easy to show that even if V_0 is just a set (maybe not a subspace), $V_0^{\perp}$ is a subspace of V_1. In our work, V_0 will always be a subspace of V_1. This allows us to decompose vectors in V_1 into the sum of two vector spaces. This decomposition is called the *direct sum*.

Definition 8.2. If V_0 is a finite-dimensional subspace of the inner product space V_1, every vector $y \in V_1$ can be decomposed uniquely as $y = v + w$, where $v \in V_0$ and $w \in V_0^{\perp}$. We say that V_1 is the *direct sum* of V_0 and $V_0^{\perp}$ and write $V_1 = V_0 \oplus V_0^{\perp}$.

Here is an example to fix these ideas. Let V_1 be the set of all piecewise-constant functions in the intervals $0 \le t < ½$ and $½ \le t < 1$, as shown in Fig. 8.6. That is, if $y(t)$ is a member of V_1 then

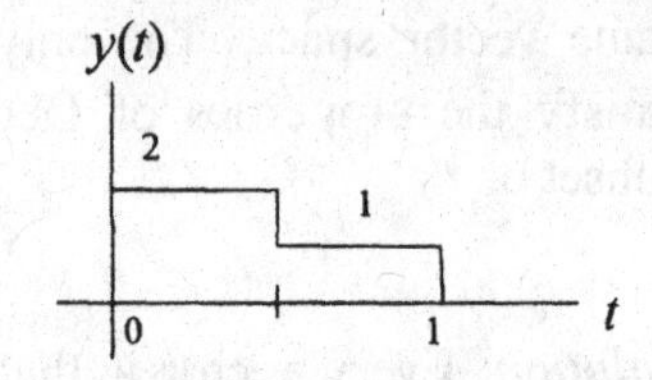

Figure 8.6. The piecewise-constant function $y(t)$.

$$y(t) = \begin{cases} k_1, & 0 \le t < \frac{1}{2} \\ k_2, & \frac{1}{2} \le t < 1 \end{cases} \tag{8.4}$$

where k_1 and k_2 are constants.

Let V_0 be the set of all constant-valued (Haar) functions in the interval $0 \le t < 1$. Figure 8.7 shows one such function, $\varphi_{00}(t) = 1$, $0 \le t < 1$. Other functions could have values of 1.5, −1, 2, and so on. In fact, $\varphi_{00}(t)$ is a basis for V_0 because any member of V_0 is simply a multiple of $\varphi_{00}(t)$. Note that $V_0 \subset V_1$, because every function in V_0 is also in V_1 (i.e., V_0 consists of functions with $k_1 = k_2$). Define the inner product for V_1 as

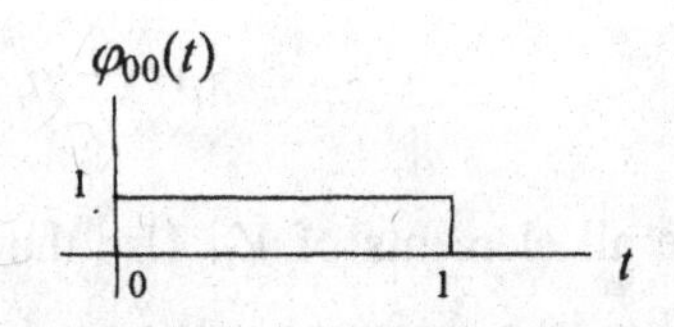

Figure 8.7. The function $\varphi_{00}(t)$.

$$\langle y_1 | y_2 \rangle = \int_0^1 y_1(t)\, y_2(t)\, dt \tag{8.5}$$

V_1 and V_0 are two vector spaces, with $V_0 \subset V_1$. Find $W_0 = V_0^{\perp}$, the orthogonal complement of V_0.

Any function with odd symmetry about the point $t = ½$ is orthogonal to every vector in V_0. However, the condition that $V_0^{\perp}$ be a subset of V_1 must also be satisfied, so our search must be restricted to piecewise-constant functions. Figure 8.8 shows a function $\psi_{00}(t)$ that satisfies both properties. It is orthogonal to every function in V_0 and it is a member of V_1. All multiples of $\psi_{00}(t)$ form the vector space $W_0 = V_0^{\perp}$.

Next, notice that the two vectors $\varphi_{00}(t)$ and $\psi_{00}(t)$ constitute a basis for V_1. That is, we can express every waveform in V_1 as a linear combination of these two vectors in one and only one way. The following example illustrates this for the waveform in Fig. 8.6.

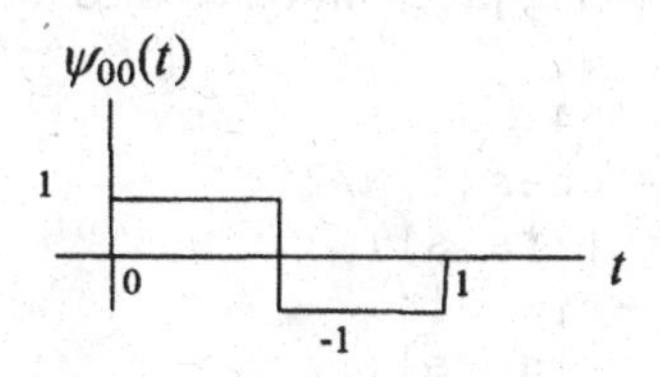

Figure 8.8. The function $\psi_{00}(t)$.

Example 8.5. Express $y(t)$ in Fig. 8.6 as a linear combination of the basis functions $\varphi_{00}(t)$ and $\psi_{00}(t)$ in Figs. 8.7 and 8.8. Assume that $k_1 = 2$ and $k_2 = 1$.

Solution: Write

$$y(t) = a_1\varphi_{00}(t) + a_2\psi_{00}(t) \tag{8.6}$$

To evaluate a_1, take the inner product of each term in this equation with $\varphi_{00}(t)$.

$$\langle y(t)|\varphi_{00}(t)\rangle = a_1\langle\varphi_{00}(t)|\varphi_{00}(t)\rangle + a_2\langle\psi_{00}(t)|\varphi_{00}(t)\rangle \tag{8.7}$$

The first inner product on the right is 1 because $\varphi_{00}(t)$ has unit norm. The second inner product is zero because $\varphi_{00}(t)$ and $\psi_{00}(t)$ are orthogonal. This gives

$$a_1 = \langle y(t)|\varphi_{00}(t)\rangle = \int_0^{1/2} 2\,dt + \int_{1/2}^1 1\,dt = \tfrac{3}{2}$$

Similarly, take the inner product of each term in Eq. 8.6 with $\psi_{00}(t)$ to get

$$a_2 = \langle y(t)|\psi_{00}(t)\rangle = \int_0^{1/2} 2\,dt - \int_{1/2}^1 1\,dt = \tfrac{1}{2}$$

giving

$$y(t) = \tfrac{3}{2}\varphi_{00}(t) + \tfrac{1}{2}\psi_{00}(t)$$

■

To implement this example in MATLAB, sample each waveform N times and use $N \times 1$ matrices instead of continuous-time functions. The following program calculates the coefficients using $N = 64$.

```
N = 64;
s = ones(1,N/2);
y = [2*s s];
p0 = [s s];
s0 = [s -s];
a1 = y*p0'/N;
a2 = y*s0'/N;
```

In this program, `p0` = $\varphi_{00}(t)$ and `s0` = $\psi_{00}(t)$. The inner product is `y*p0'` because `y` and `p0` are row vectors. Dividing by the number of samples normalizes the inner products. The other statements should be self-explanatory.

Example 8.6. Let V_1 be the set of all discrete-time signals of length 2. For example, $v(n) = \{2, -1\}$ is one signal in the set V_1. Let V_0 be the subset of V_1 where the two signal values are equal. Thus $\varphi_{00}(n) = \{1, 1\}$ is a member of V_0.

(a) Show that V_0 is a subspace of V_1.
(b) Find the orthogonal complement $W_0 = V_0^{\perp}$. Select as $\psi_{00}(n)$ one member of W_0.
(c) Show that $\varphi_{00}(n)$ and $\psi_{00}(n)$ together constitute a basis for V_1.

Solution: (a) V_0 is a subspace of V_1 because the zero vector $\{0, 0\}$ is a member of V_0, and linear combinations of vectors in V_0 are also in V_1. That is, if $v_1(n) = \{a, a\}$ and $v_2(n) = \{b, b\}$, then $v_1(n) + v_2(n)$ also has the same value for the two signal values, and is therefore a member of V_0. V_0 automatically inherits the other properties of a vector space since V_1 is a vector space and V_0 is a subset of V_1
(b) The orthogonal complement $W_0 = V_0^{\perp}$ is the set of all vectors in V_1 that are orthogonal to every vector in V_0. This means that there are two requirements for a signal $\psi_{00}(n)$ to be a member of W_0: (1) $\psi_{00}(n)$ must be in V_1; (2) $\psi_{00}(n)$ must be orthogonal to all members of V_0. Thus, $\psi_{00}(n)$ must have length 2 and the inner product between $\psi_{00}(n)$ and every member of V_0 must be zero. Both conditions are satisfied by $\psi_{00}(n) = \{1, -1\}$.

(c) $\varphi_{00}(n)$ and $\psi_{00}(n)$ constitute a basis for V_1 because every signal of length 2 can be expressed as a linear combination of these two signals. That is, for an arbitrary signal $v(n)$ of length 2, unique scalars a_1 and a_2 can be found such that

$$v(n) = a_1\varphi_{00}(n) + a_2\psi_{00}(n)$$

Notice that this also shows that every vector in V_1 is in either V_0 or W_0, meaning that

$$V_1 = V_0 \oplus W_0.$$

■

Drill 8.2. Let $v(t) = -u(t) + 3u(t-0.5) - 2u(t-1)$, where $u(t)$ is the familiar unit step function. Express $v(t)$ as a linear combination of the basis functions in Example 8.5.

Answer: $v(t) = \frac{1}{2}\varphi_{00}(t) - \frac{3}{2}\psi_{00}(t).$

Example 8.7. Find the coefficients to approximate $y(t) = 1 + \sin(2\pi t)$ by the functions $\varphi_{00}(t)$ and $\psi_{00}(t)$ in Figs. 8.7 and 8.8.

Solution: $a_1 = \langle y(t) | \varphi_{00}(t)\rangle = \int_0^1 [1+\sin(2\pi t)]\,dt = 1$

$$a_2 = \langle y(t) | \psi_{00}(t)\rangle = \int_0^{1/2} [1+\sin(2\pi t)]\,dt - \int_{1/2}^1 [1+\sin(2\pi t)]\,dt = \frac{2}{\pi}$$

Therefore,

$$y(t) \approx \varphi_{00}(t) + (2/\pi)\psi_{00}(t)$$

The following MATLAB program produces the picture of this in Fig. 8.9. The rectangular waveform is the sum $\varphi_{00}(t) + (2/\pi)\psi_{00}(t)$, which approximates the sinusoid $y(t)$ as closely as possible.

```
N = 64;
s = ones(1,N/2);
t = linspace(0,1,N);
y = 1+sin(2*pi*t);
p0 = [s s];
s0 = [s -s];
```

```
a1 = y*p0'/N;
a2 = y*s0'/N;
v = a1*p0 + a2*s0;
plot(t,y,t,v)
```

This program calculates the values of a_1 and a_2 instead of using the true values. This introduces error since we are using samples of $y(t)$. The true value of a_2 is $2/\pi = 0.6366$. The value derived in this program is $a_2 = 0.6265$. If N is increased to 256 in the program, then $a_2 = 0.6341$. Therefore, we should use a large value of N in our programs to improve accuracy.

■

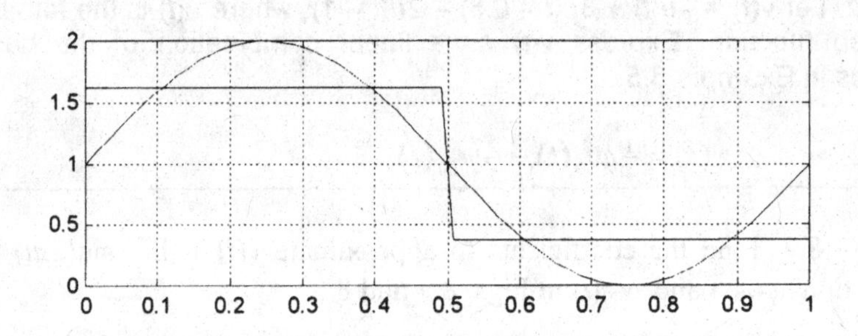

Figure 8.9. Approximating a sinusoid by $\varphi_{00}(t)+\frac{2}{\pi}\psi_{00}(t)$.

Here is a review of some notation relating to waveforms. Let $v(t)$ be the waveform in Fig. 8.10a. Then $v(2t)$, $v[2(t-1)]$, and $v(2t-1)$ have the shapes shown in Fig. 8.10b, c, and d, respectively.

Drill 8.3. Suppose that $v(t) = e^{-t}u(t)$. Plot $v(t)$, $v(2t)$, $v[2(t-1)]$, $v(2t-1)$.

Answer: Figure 8.11.

Notation is important to clear thinking, and the old notation of $V_1 = V_0 \oplus V_0^{\perp}$ is awkward. The notation should be fixed in preparation for expanding on the themes of subspaces, direct sums, and orthogonal complements. The last example used good notation. Let

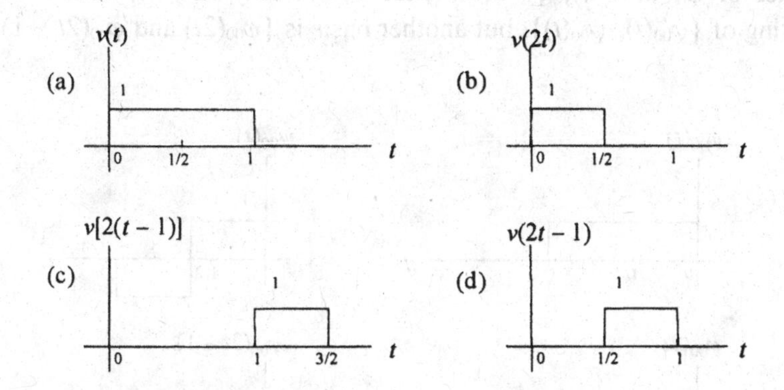

Figure 8.10. Dilations and translations of $v(t)$.

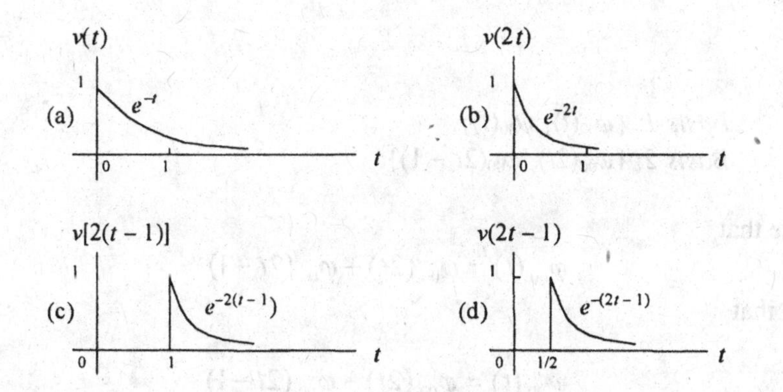

Fig. 8.11. Drill 8.3 answer.

$$W_0 = V_0^{\perp}$$

Then we can write

$$V_1 = V_0 \oplus W_0$$

Refer to Fig. 8.12. If $\varphi_{00}(t)$ and $\psi_{00}(t)$ form a basis for V_1, then V_0 consists of multiples of $\varphi_{00}(t)$, W_0 consists of multiples of $\psi_{00}(t)$, and V_1 consists of multiples of $\varphi_{00}(2t - k)$, $k = 0,1$. That is, not only does V_1 have a basis consisting of $\{\varphi_{00}(t), \psi_{00}(t)\}$, but another basis is $\{\varphi_{00}(2t)$ and $\varphi_{00}(2t-1)\}$.

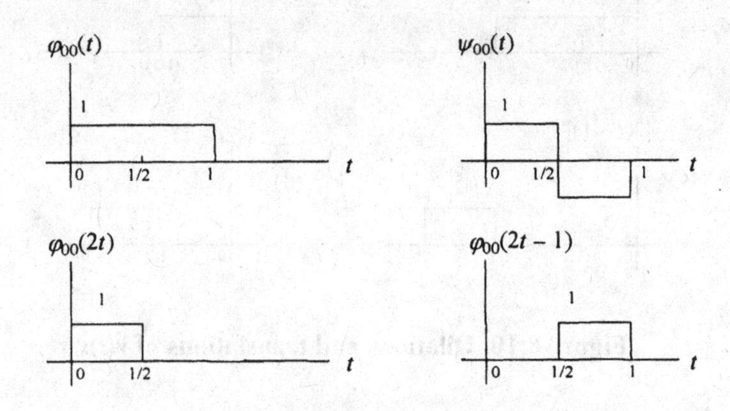

Figure 8.12. Bases 1 and 2.

Basis 1: $\{\varphi_{00}(t), \psi_{00}(t)\}$.
Basis 2: $\{\varphi_{00}(2t), \varphi_{00}(2t-1)\}$.

Note that

$$\varphi_{00}(t) = \varphi_{00}(2t) + \varphi_{00}(2t-1)$$

and that

$$\psi_{00}(t) = \varphi_{00}(2t) - \varphi_{00}(2t-1)$$

Since $\{\varphi_{00}(t), \psi_{00}(t)\}$ can be expressed as a linear combination of the $\{\varphi_{00}(2t), \varphi_{00}(2t-1)\}$ vectors, then $\{\varphi_{00}(2t), \varphi_{00}(2t-1)\}$ must also be a basis for V_1.

To explore this second basis a bit more, let

$$\varphi_{jk}(t) = 2^{j/2}\varphi_{00}(2^j t - k) \tag{8.8}$$

where the $2^{j/2}$ factor normalizes each function to have unit energy (unit norm). This notation means that

$$j=0,\ k=0,\ 2^0\varphi_{00}(2^0t-0)=\varphi_{00}(t)$$
$$j=1,\ k=0,\ 2^{1/2}\varphi_{00}(2^1t-0)=\sqrt{2}\varphi_{00}(2t)$$
$$j=1,\ k=1,\ 2^{1/2}\varphi_{00}(2^1t-1)=\sqrt{2}\varphi_{00}(2t-1)$$
$$j=2,\ k=0,\ 2^1\varphi_{00}(2^2t-0)=2\varphi_{00}(4t)$$
etc.

where k ranges from 0 to $2^j - 1$. Ideally, j starts at 0 and goes to ∞. In practice, the upper limit on j is determined by the sampling rate. Practical calculation of wavelet coefficients starts at the other end. Instead of starting with $j = 0$ and working our way up to the maximum value of j, we begin at the finest resolution and work our way down. More on this later. As shown in Fig. 8.13 for basis 2,

$$\varphi_{10}(t)=\sqrt{2}\varphi_{00}(2t) \quad \text{and} \quad \varphi_{11}(t)=\sqrt{2}\varphi_{00}(2t-1)$$

So far the discussion has been limited to V_0 and V_1. To define a further refinement of the interval $0 \le t < 1$, let V_2 be the set of all piecewise constant functions in the intervals $0 \le t < ¼$, $¼ \le t < ½$, $½ \le t < ¾$, and $¾ \le t < 1$. These are the functions φ_{2k} shown in Fig. 8.13. From Eq. 8.8,

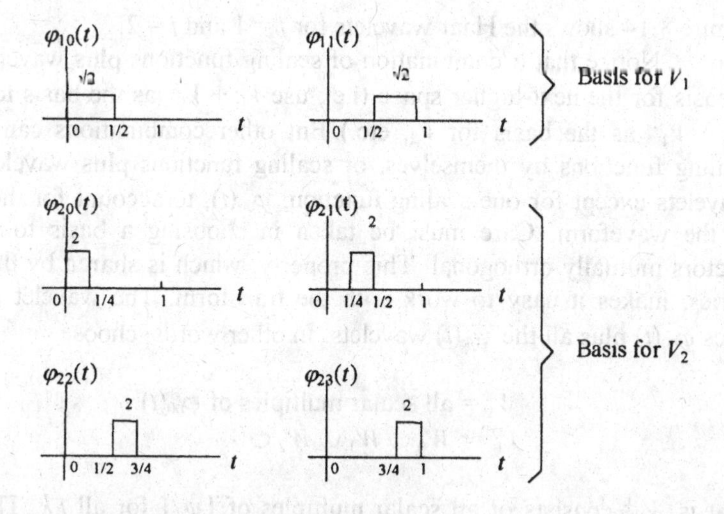

Figure 8.13. Bases for V_1 and V_2.

$$\varphi_{2k}(t) = 2\varphi_{00}(4t - k), \qquad k = 0, 1, 2, 3$$

The $\varphi_{1k}(t)$ functions form a basis for V_1, and the $\varphi_{2k}(t)$ functions form a basis for V_2. [Note that the $\varphi_{2k}(t)$ functions *do not* form a basis for V_1, although each vector in V_1 can be expressed as a linear combination of the $\varphi_{2k}(t)$ functions. Why not? Because there are too many vectors in the set $\{\varphi_{2k}(t)\}$. For this to be a basis, there should be only two functions in the set.]

The $\varphi_{1k}(t)$ functions form a basis for V_1, the $\varphi_{2k}(t)$ functions form a basis for V_2, and this could keep going. This generates a hierarchy of vector spaces defined by their basis functions in Eq. 8.8.

$$V_0 \subset V_1 \subset V_2 \subset \cdots \subset L^2 \tag{8.9}$$

where L^2 is the set of all energy signals.

The functions $\{\varphi_{jk}\}$ are called *scaling functions*, and the functions $\{\psi_{jk}\}$ are called *wavelets*. Scaling functions form a basis for V_j. Wavelets form a basis for $W_j = V_j^{\perp}$. Similar to the scaling functions, they are defined by

$$\psi_{jk}(t) = 2^{j/2}\psi_{00}(2^j t - k) \tag{8.10}$$

Figure 8.14 shows the Haar wavelets for $j = 1$ and $j = 2$.

Notice that a combination of scaling functions plus wavelets forms a basis for the next-higher space (i.e., use $V_0 + V_0^{\perp}$ as the basis for V_1, use $V_1 + V_1^{\perp}$ as the basis for V_2, etc.) But other combinations can be used: scaling functions by themselves, or scaling functions plus wavelets, or all wavelets except for one scaling function, $\varphi_{00}(t)$, to account for the dc level in the waveform. Care must be taken in choosing a basis to make the vectors mutually orthogonal. This property, which is shared by the Fourier series, makes it easy to work with the transform. The wavelet transform uses $\varphi_{00}(t)$ plus all the $\psi_{jk}(t)$ wavelets. In other words, choose

$$V_0 = \text{all scalar multiples of } \varphi_{00}(t).$$
$$V_0^{\perp} = W_0 \cup W_1 \cup W_2 \cup \cdots$$

that is, $V_0^{\perp}$ consists of all scalar multiples of $\{\psi_{jk}\}$ for all j,k. This makes every vector in the basis orthogonal to all other basis vectors at all resolutions. To see how this works, consider the following:

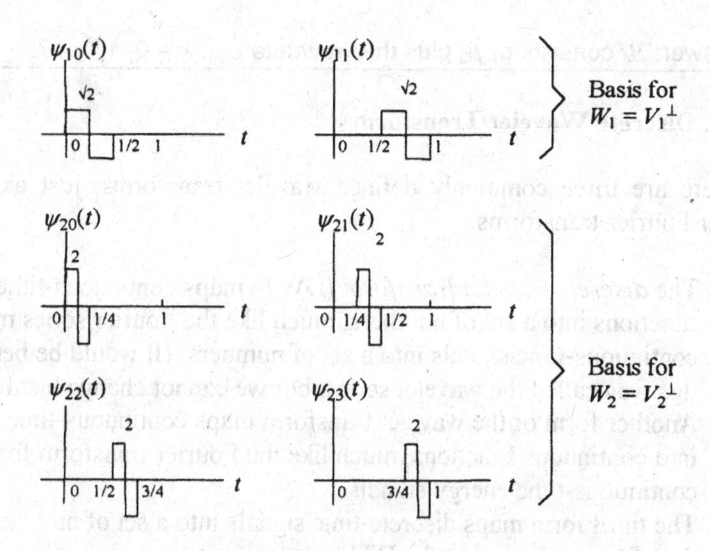

Figure 8.14. Haar wavelets for $j = 1$ and $j = 2$.

1. This scheme generates a hierarchy of vector spaces.

$$V_0 \subset V_1 \subset V_2 \subset \cdots \subset L^2$$

2. Each vector space V_i can be decomposed into $V_i = V_{i\text{-}1} \oplus W_{i\text{-}1}$, where $W_{i\text{-}1}$ is the orthogonal complement $V_{i\text{-}1}^{\perp}$. This means that V_i has a basis composed of the basis for $V_{i\text{-}1}$ combined with the basis for $W_{i\text{-}1}$. For example, V_2 has the basis

$$\beta_2 = \{\varphi_{00}, \psi_{00}, \psi_{10}, \psi_{11}\}$$

V_3 has the basis consisting of β_2 plus the wavelets that constitute a basis for W_2. Thus

$$\beta_3 = \{\varphi_{00}, \psi_{00}, \psi_{10}, \psi_{11}, \psi_{20}, \psi_{21}, \psi_{22}, \psi_{23}\}$$

These examples illustrate the scheme for constructing the bases for higher resolutions.

Drill 8.4. Construct a wavelet basis for V_4.

Answer: β_4 consists of β_3 plus the wavelets ψ_{3k}, $k = 0, 1, \ldots, 7$.

8.2. Discrete Wavelet Transform

There are three commonly defined wavelet transforms, just as there are four Fourier transforms.

1. The *discrete wavelet transform* (DWT) maps continuous-time functions into a set of numbers, much like the Fourier series maps continuous-time signals into a set of numbers. (It would be better if this was called the wavelet series, but we cannot change history.)
2. Another form of the wavelet transform maps continuous-time signals into continuous functions, much like the Fourier transform for continuous-time energy signals.
3. The third form maps discrete-time signals into a set of numbers, and is therefore analogous to the DTFS.

This section concentrates on the DWT. It maps continuous-time functions into numbers. The forward (analysis) equations are given by

$$\begin{aligned} c_{jk} &= \langle v(t) | \varphi_{jk}(t) \rangle \\ d_{jk} &= \langle v(t) | \psi_{jk}(t) \rangle \end{aligned} \tag{8.11}$$

The inverse (synthesis) equation is given by

$$v(t) = \sum_{k=-\infty}^{\infty} c_{Jk} \varphi_{Jk}(t) + \sum_{j=J}^{\infty} \sum_{k=-\infty}^{\infty} d_{jk} \psi_{jk}(t) \tag{8.12}$$

where J is the starting index (usually, $J = 0$). The analysis equations decompose (analyze) a given waveform $v(t)$ into its constituent components (building blocks) c_{jk} and d_{jk}. Figure 8.15a pictures this operation. The synthesis equation builds (synthesize) the function $v(t)$ from its components c_{jk} and d_{jk}. Figure 8.15b shows this operation.

Example 8.8. Find the wavelet transform of the function in Fig. 8.16. Use the four functions in Fig. 8.17 for the basis.

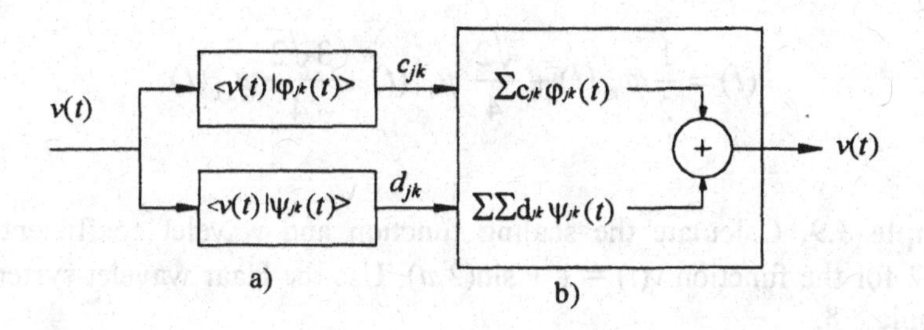

Figure 8.15. Forward and inverse wavelet transforms.

Solution: The forward (analysis) equations (Eq. 8.8) give

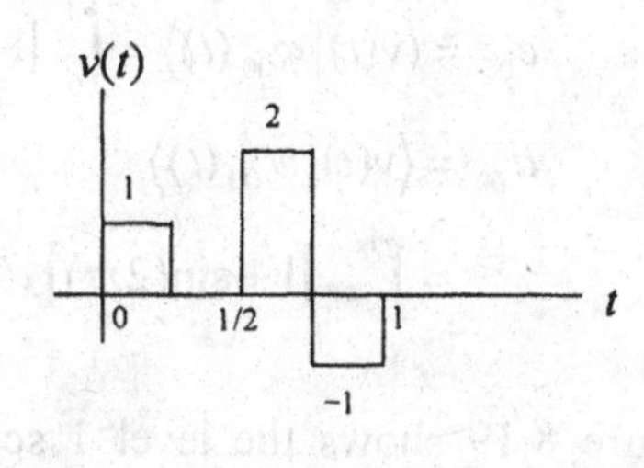

Figure 8.16. Function *v*(*t*).

$$c_{00} = \langle v(t) | \varphi_{00}(t) \rangle = \frac{1}{4} + \frac{2}{4} - \frac{1}{4} = \frac{1}{2}$$

$$d_{00} = \langle v(t) | \psi_{00}(t) \rangle = \frac{1}{4} - \frac{2}{4} + \frac{1}{4} = 0$$

$$d_{10} = \langle v(t) | \psi_{10}(t) \rangle = \frac{\sqrt{2}}{4}$$

$$d_{11} = \langle v(t) | \psi_{11}(t) \rangle = \frac{2\sqrt{2}}{4} + \frac{\sqrt{2}}{4} = \frac{3\sqrt{2}}{4}$$

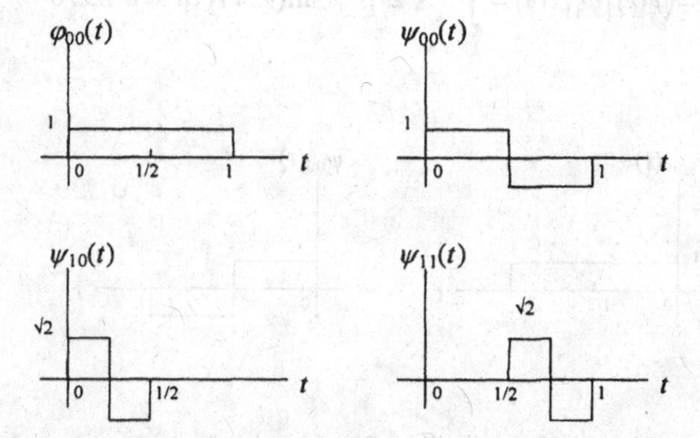

Figure 8.17. Basis functions for Example 8.8.

giving

$$v(t) = \frac{1}{2}\varphi_{00}(t) + \frac{\sqrt{2}}{4}\psi_{10}(t) + \frac{3\sqrt{2}}{4}\psi_{11}(t)$$

■

Example 8.9. Calculate the scaling function and wavelet coefficients to level 2 for the function $v(t) = 1 + \sin(2\pi t)$. Use the Haar wavelet system in Example 8.8.

Solution: Figure 8.18 shows the scaling function and mother wavelet. The zero-level coefficients are given by

$$c_{00} = \langle v(t) | \varphi_{00}(t) \rangle = \int_0^1 [1 + \sin(2\pi t)]\, dt = 1$$

$$d_{00} = \langle v(t) | \psi_{00}(t) \rangle$$

$$= \int_0^{1/2} [1 + \sin(2\pi t)]\, dt - \int_{1/2}^1 [1 + \sin(2\pi t)]\, dt = 0.6366$$

Figure 8.19 shows the level 1 scaling and wavelet functions. Their inner product with $v(t)$ gives

$$c_{10} = \langle v(t) | \varphi_{10}(t) \rangle = \int_0^{1/2} \sqrt{2}\, [1 + \sin(2\pi t)]\, dt = 1.1573$$

$$c_{11} = \langle v(t) | \varphi_{11}(t) \rangle = \int_{1/2}^1 \sqrt{2}\, [1 + \sin(2\pi t)]\, dt = 0.2570$$

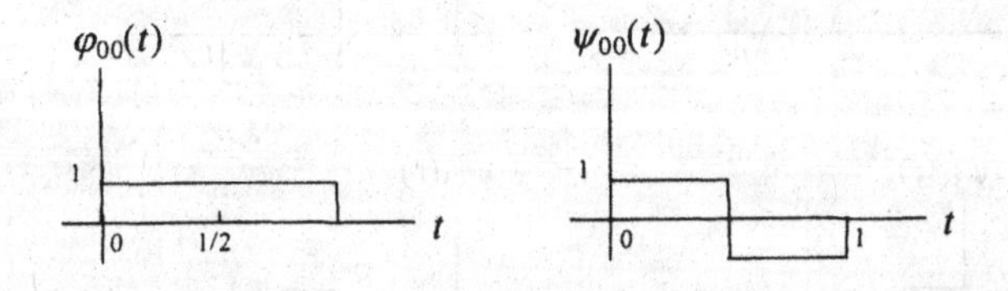

Figure 8.18. Haar scaling function and mother wavelet.

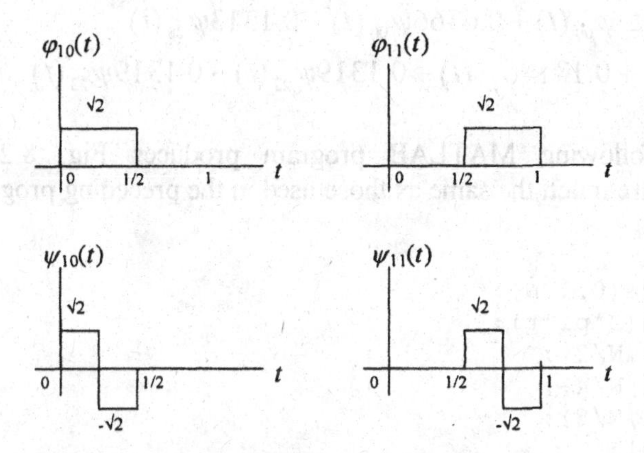

Fig. 8.19. The level 1 scaling and wavelet functions.

$$d_{10} = \langle v(t) | \psi_{10}(t) \rangle$$
$$= \int_0^{1/4} \sqrt{2}\left[1+\sin(2\pi t)\right] dt - \int_{1/4}^{1/2} \sqrt{2}\left[1+\sin(2\pi t)\right] dt = 0$$

$$d_{11} = \langle v(t) | \psi_{11}(t) \rangle$$
$$= \int_{1/2}^{3/4} \sqrt{2}\left[1+\sin(2\pi t)\right] dt - \int_{3/4}^{1} \sqrt{2}\left[1+\sin(2\pi t)\right] dt = 0$$

In a similar manner, the second level coefficients are given by

$$c_2 = \begin{bmatrix} c_{20} \\ c_{21} \\ c_{22} \\ c_{23} \end{bmatrix} = \begin{bmatrix} 0.8183 \\ 0.8183 \\ 0.1817 \\ 0.1817 \end{bmatrix} \qquad d_2 = \begin{bmatrix} d_{20} \\ d_{21} \\ d_{22} \\ d_{23} \end{bmatrix} = \begin{bmatrix} -0.1319 \\ 0.1319 \\ 0.1319 \\ -0.1319 \end{bmatrix}$$

This gives the series

$$v(t) = 1 + \sin(2\pi t) \cong c_{00}\varphi_{00}(t) + d_{00}\psi_{00}(t) + d_{10}\psi_{10}(t) + d_{11}\psi_{11}(t)$$
$$+ d_{20}\psi_{20}(t) + d_{21}\psi_{21}(t) + d_{22}\psi_{22}(t) + d_{23}\psi_{23}(t)$$

or

$$v(t) \cong \varphi_{00}(t) + 0.6366\psi_{00}(t) - 0.1313\psi_{20}(t)$$
$$+0.1319\psi_{21}(t) + 0.1319\psi_{22}(t) - 0.1319\psi_{23}(t)$$

The following MATLAB program produces Fig. 8.20. The symbols used are much the same as those used in the preceding program.

```
N = 16384;
t = linspace(0,1,N);
y = 1 + sin(2*pi*t);
e0 = ones(1,N/2);
e1 = ones(1,N/4);
e2 = ones(1,N/8);
p0 = [e0 e0];
s0 = [e0 -e0];
c0 = y*p0'/N;
d00 = y*s0'/N;
s10 = sqrt(2)*[e1 -e1 zeros(1,N/2)];
s11 = sqrt(2)*[zeros(1,N/2) e1 -e1];
d10 = y*s10'/N;
d11 = y*s11'/N;
s20 = 2*[e2 -e2 zeros(1,3*N/4)];
s21 = 2*[zeros(1,N/4) e2 -e2 zeros(1,N/2)];
s22 = 2*[zeros(1,N/2) e2 -e2 zeros(1,N/4)];
s23 = 2*[zeros(1,3*N/4) e2 -e2];
d20 = y*s20'/N;
d21 = y*s21'/N;
d22 = y*s22'/N;
d23 = y*s23'/N;
v = c0*p0 + d00*s0 +d10*s10 + d11*s11 + d20*s20...
+ d21*s21 + d22*s22 + d23*s23
subplot(2,1,1)
plot(t,y,t,v)
grid on
```

■

To this point only Haar wavelets have been discussed, those with the scaling function in Fig. 8.12. Before proceeding further with examples of other wavelets, let us put together some concepts that are implied in our discussion so far.

1. The construction of wavelets uses scaling functions $\varphi(t)$ that can be written as linear combinations of $\varphi(2t - k)$, the half-scaled and $k/2$ translated versions of $\varphi(t)$. This linear combination is

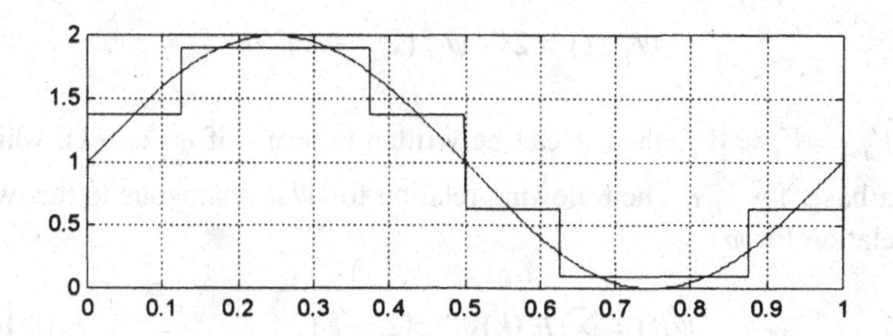

Figure 8.20. Approximating $v(t) = 1 + \sin(2\pi t)$ with $j = 2$.

$$\varphi(t) = \sum_k h_0(k)\sqrt{2}\,\varphi(2t-k) \tag{8.13}$$

where the terms $\sqrt{2}\,h_0(k)$ are the coefficients that relate the $\varphi(2t-k)$ function to $\varphi(t)$. This is called the *two-scale relation* for φ, and the coefficients $\sqrt{2}\,h_0(k)$ are called the *two-scale sequence* of φ. (Some examples follow this list to illustrate Eq. 8.13.)

2. The subspace V_0 is spanned by $\varphi_0(t)$ and its integer translates.
3. The functions in Eq. 8.5,

$$\varphi_{jk}(t) = 2^{j/2}\varphi_0(2^j t - k)$$

span the space V_j. The two-scale relationship in Eq. 8.13 generates a nested sequence of subspaces:

$$\cdots V_{-1} \subset V_0 \subset V_1 \subset V_2 \subset \cdots$$
$$\leftarrow \text{coarser} \qquad \text{finer} \rightarrow$$

This is called *multiresolution analysis* (MRA).

4. The space V_j and its orthogonal complement W_j are both subsets of V_{j+1}. That is,

$$V_{j+1} = V_j \oplus W_j$$

Given a scaling function φ in V_j, the basic tenet of MRA is that there exists another function ψ in W_j called the *wavelet*, where (Eq. 8.10)

$$\psi_{jk}(t) = 2^{j/2}\psi_{00}(2^j t - k)$$

Since $V_{j+1} = V_j \oplus W_j$, then ψ can be written in terms of $\varphi(2t - k)$, which forms a basis for V_{j+1}. The following relation for ψ is analogous to the two-scale relation for φ :

$$\psi(t) = \sum_k h_1(k)\sqrt{2}\,\varphi(2t - k) \tag{8.14}$$

This is the two-scale relation for wavelets.

Example 8.10. Given the two functions $\varphi(2t)$ and $\varphi(2t - 1)$ in Fig. 8.21, and given the filters h_0 and h_1 as

$$h_0(k) = \left\{\tfrac{1}{\sqrt{2}} \quad \tfrac{1}{\sqrt{2}}\right\}$$

$$h_1(k) = \left\{\tfrac{1}{\sqrt{2}} \quad -\tfrac{1}{\sqrt{2}}\right\}$$

Use Eqs. 8.13 and 8.14 to find the scaling function $\varphi(t)$ and the "mother wavelet" $\psi(t)$.

Solution: $\varphi(t) = \sum_k h_0(k)\sqrt{2}\,\varphi(2t - k)$

$$= \underbrace{\tfrac{1}{\sqrt{2}}\sqrt{2}\,\varphi(2t)}_{k=0} + \underbrace{\tfrac{1}{\sqrt{2}}\sqrt{2}\,\varphi(2t-1)}_{k=1} = \varphi(2t) + \varphi(2t - 1)$$

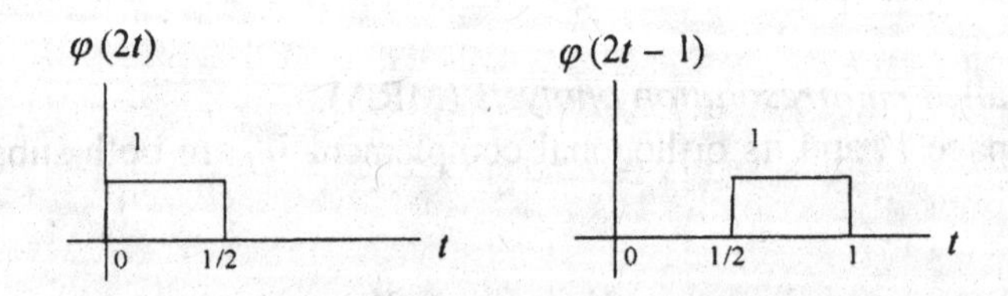

Figure 8.21

This gives $\varphi(t)$, shown in Fig. 8.22a. Also,

$$\psi(t) = \sum_k h_1(k)\sqrt{2}\,\varphi(2t-k)$$
$$= \underbrace{\tfrac{1}{\sqrt{2}}\sqrt{2}\,\varphi(2t)}_{k=0} - \underbrace{\tfrac{1}{\sqrt{2}}\sqrt{2}\,\varphi(2t-1)}_{k=1} = \varphi(2t) - \varphi(2t-1)$$

This gives $\psi(t)$, shown in Fig. 8.22b.

■

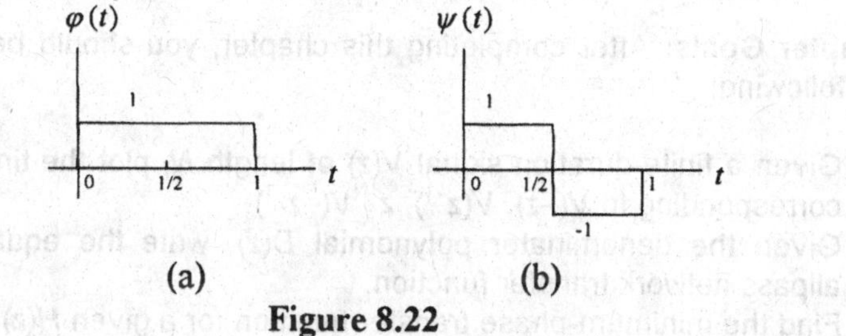

Figure 8.22

Notice that Equations 8.13 and 8.14 hold at any scale. A fixed number of coefficients $h_0(k)$ and $h_1(k)$ relate the scaling functions and wavelets at one resolution to the scaling functions at the next lower resolution. This, plus the fact that wavelets form an orthonormal basis, will in Chapter 10 lead to some useful relations that allow easy calculation of the wavelet coefficients at all resolutions.

Chapter 9
Quadrature Mirror Filters

This section introduces several concepts that are needed for the study of wavelet transforms. Among these is quadrature mirror filters (QMFs), allpass networks, mirror-image singularities, and minimum, maximum, and linear phase networks. The filter banks used in wavelet analysis are made of QMFs. The concepts of minimum phase, maximum phase, and linear phase filters are needed to understand how QMFs are used in wavelet analysis.

Chapter Goals: After completing this chapter, you should be able to do the following:

- Given a finite-duration signal $V(z)$ of length N, plot the time functions corresponding to $V(-z)$, $V(z^{-1})$, $z^{-N}V(-z^{-1})$.
- Given the denominator polynomial $D(z)$, write the equation for an allpass network transfer function.
- Find the minimum-phase transfer function for a given $H(z)$.
- Given $H_0(z)$, find the corresponding QMF and conjugate QMF.
- Calculate the slope of the phase for a symmetric finite-impulse response filter.
- Determine if a given two-channel filter bank is a perfect reconstruction filter.

9.1. Allpass Networks

A finite-impulse response (FIR) filter has an impulse response that lasts for a definite time. That is, the impulse response $h(n)$ is zero after some definite time. The structure in Fig. 9.1 has an impulse response $h_0(n) = \{1, 1\}$, which is shorthand notation for

$$h_0(n) = \delta(n) + \delta(n-1) \qquad (9.1)$$

The input $v(n) = \delta(n)$ is multiplied by 1 and summed with a delayed version of $v(n)$ to produce the sum in Eq. 9.1.

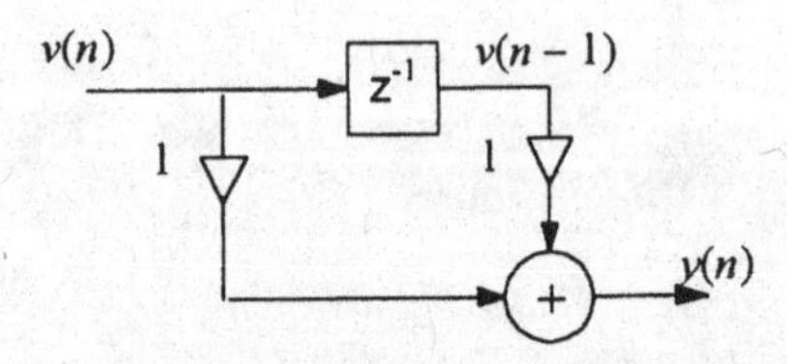

Figure 9.1. Low-pass filter.

The structure in Fig. 9.1 is a low-pass filter. To see this, take the z transform of the impulse response $h_0(n)$ to obtain the transfer function $H_0(z)$.

$$H_0(z) = 1 + z^{-1} = \frac{z+1}{z} \tag{9.2}$$

This has one pole at $z = 0$, and one zero at $z = -1$, as shown in Fig. 9.2. Figure 9.3 shows the z-plane magnitude (the rubber sheet) for this pole-zero configuration.

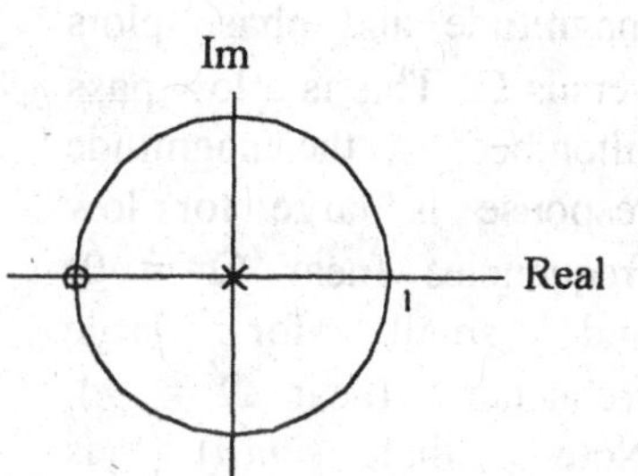

Figure 9.2. Pole-zero plot.

To find the frequency response $H_0(\Omega) = H_0(e^{j\Omega})$, substitute $z = e^{j\Omega}$ into Eq. 9.2.

$$H_0(\Omega) = 1 + e^{-j\Omega} \tag{9.3}$$

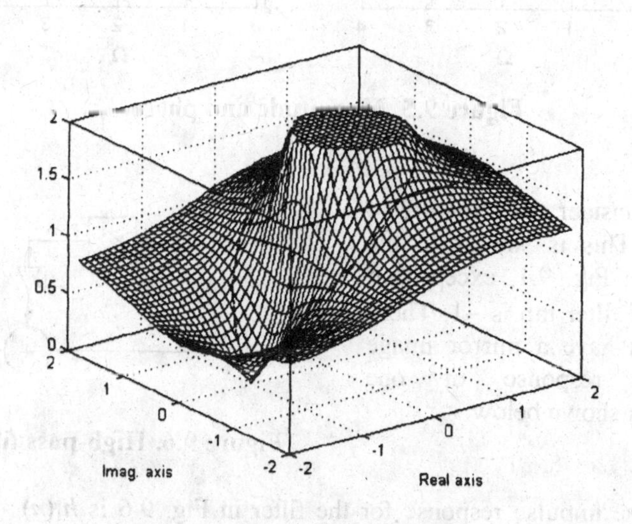

Figure 9.3. Rubber sheet.

Figure 9.4 shows the magnitude of this function around the unit circle, and Fig. 9.5 shows the magnitude and phase plots versus Ω. This is a low-pass filter because the magnitude response is large for low frequencies (near $\Omega = 0$) and small for large frequencies (near $\Omega = \pi$). Note that $h(n)$ is symmetrical about its midpoint at $n = 1.5$, and the phase in Fig. 9.5 is linear.

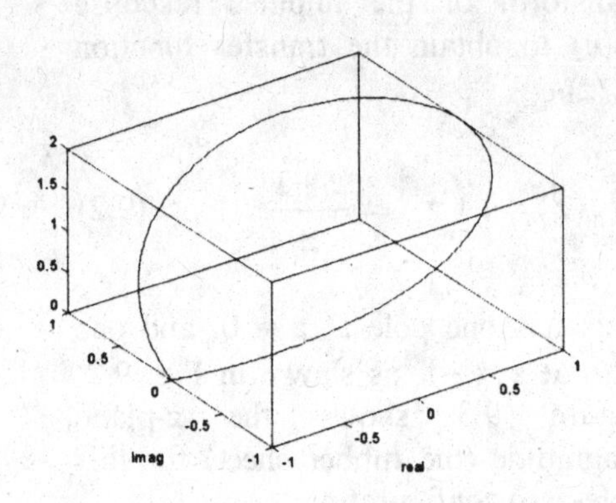

Figure 9.4. Unit circle magnitude.

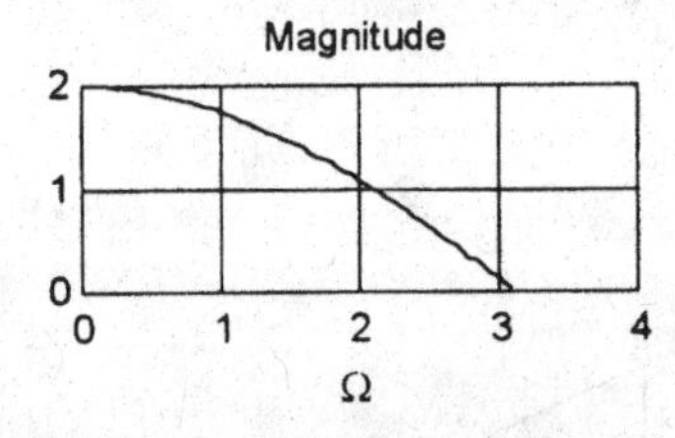

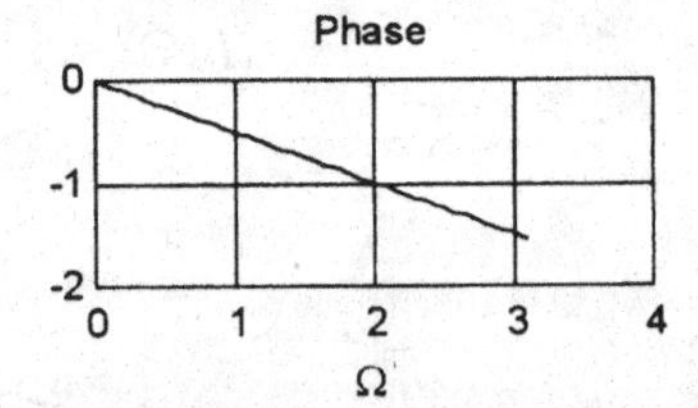

Figure 9.5. Magnitude and phase.

Consider the structure in Fig. 9.6. This is similar to the example in Fig. 9.1, except that the second filter tap is −1. These two filters have a mirror image frequency response of one another, as shown below.

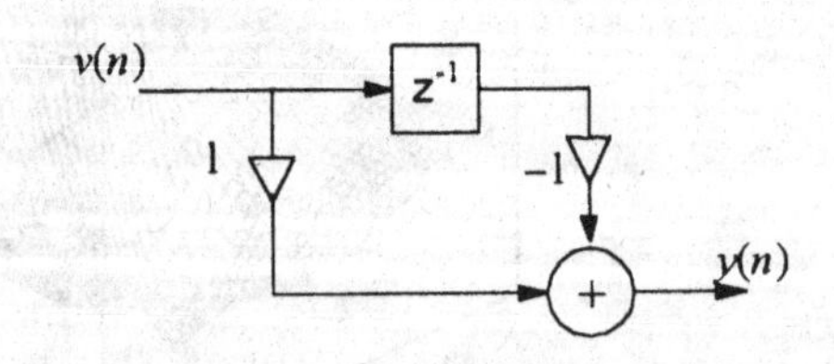

Figure 9.6. High-pass filter.

The impulse response for the filter in Fig. 9.6 is $h_1(n) = \{1, -1\}$. The z transform gives the transfer function,

$$H_1(z) = 1 - z^{-1} = \frac{z-1}{z} \qquad (9.4)$$

This is a high-pass filter. Figure 9.7 shows the pole-zero plot, and Fig. 9.8 shows the magnitude in the z-plane. The response versus frequency Ω is given by

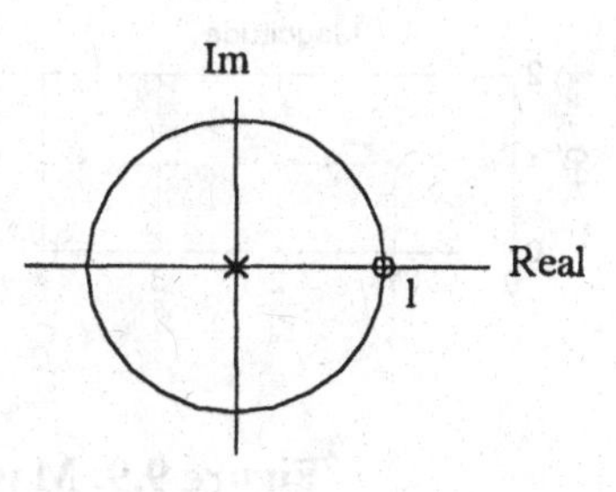

Figure 9.7. Pole-zero plot.

$$H_1(e^{j\Omega}) = H_1(\Omega) = 1 - e^{-j\Omega} \qquad (9.5)$$

Figure 9.9 shows the magnitude and phase versus Ω. This is a high-pass filter because the magnitude response is small for low frequencies and large for high frequencies around $\Omega = \pi$.

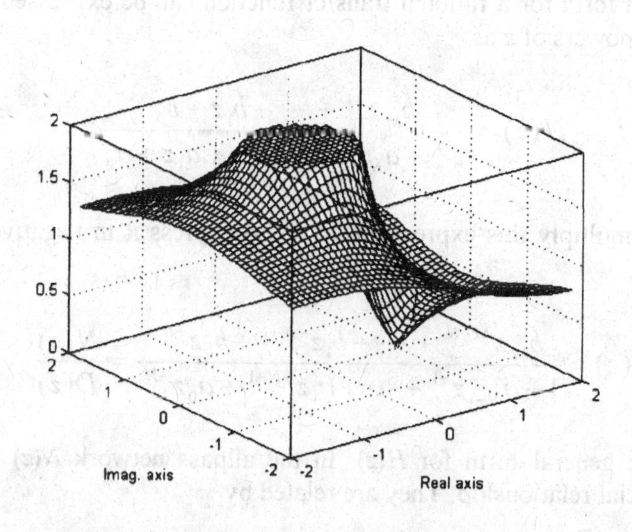

Figure 9.8. Rubber sheet.

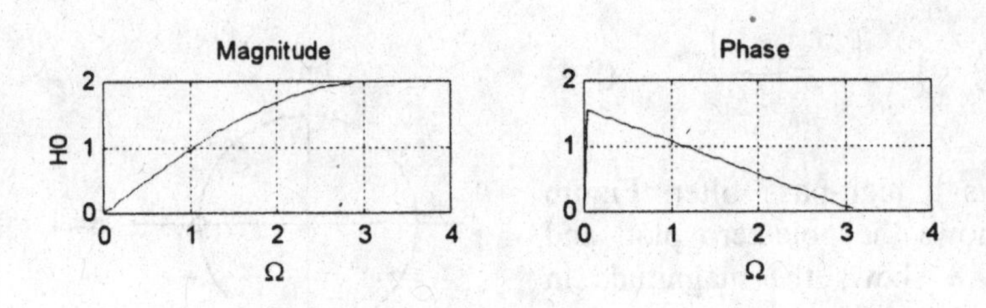

Figure 9.9. Magnitude and phase.

In comparing the diagrams for these two filters, note that they are mirror images of each other in some sense. This concept of mirror image filters will be so important to our study of wavelets that we will pause here to introduce the subject in some detail. The concept of mirror symmetry comes about in the following way. An *allpass network* has equal magnitude response at all frequencies, that is,

$$|H(\Omega)|^2 = 1 \quad \text{for all } \Omega \tag{9.6}$$

The general form for a rational transfer function can be expressed in terms of positive powers of z as

$$H(z) = \frac{b_M z^M + \cdots + b_1 z + b_0}{z^N + a_{N-1} z^{N-1} + \cdots + a_1 z + a_0} \tag{9.7a}$$

or we can multiply this expression by z^{-N} to express it in negative powers of z.

$$H(z) = \frac{b_M z^{M-N} + \cdots + b_1 z^{-N+1} + b_0 z^{-N}}{1 + a_{N-1} z^{-1} + \cdots + a_1 z^{-N+1} + a_0 z^{-N}} = \frac{N(z)}{D(z)} \tag{9.7b}$$

This is the general form for $H(z)$. In the allpass network $N(z)$ and $D(z)$ have a special relationship. They are related by

$$N(z) = z^{-N} D(z^{-1}) \tag{9.8}$$

This implies first that $N(z)$ and $D(z)$ have the same order, and second that the coefficient sequences are in reverse order. For example, if $D(z)$ is given by

$$D(z) = 1 + 2z^{-1} + 3z^{-2} + 4z^{-3} + 5z^{-4}$$

then

$$D(z^{-1}) = 1 + 2z + 3z^{2} + 4z^{3} + 5z^{4}$$

and

$$N(z) = z^{-4}D(z^{-1}) = 5 + 4z^{-1} + 3z^{-2} + 2z^{-3} + z^{-4}$$

giving

$$H(z) = \frac{5 + 4z^{-1} + 3z^{-2} + 2z^{-3} + z^{-4}}{1 + 2z^{-1} + 3z^{-2} + 4z^{-3} + 5z^{-4}}$$

The following drill problem is designed to help you become familiar with this notation.

Drill 9.1. The time-reversal property of z transforms says that if $v(n) \leftrightarrow V(z)$, then $v(-n) \leftrightarrow V(z^{-1})$. To see what this means, do the following for the waveforms in Fig. 9.10.
(a) Find the *z* transform of $v_1(n)$.
(b) Find the z transform of $v_2(n)$ and note that it can be expressed as $V_1(z^{-1})$.
(c) Find the z transform of $v_3(n)$ and note that it can be expressed as $V_1(-z)$.
(d) Plot the time function corresponding to $V_1(-z^{-1})$.

Answer: (a) $V_1(z) = 1 + 2z^{-1} + 3z^{-2} + 4z^{-3} + 5z^{-4}$.
(d) Last diagram in Fig. 9.10.

Using some of the results of Drill 9.1, note that $D(z^{-1})$ in Eq. 9.8 reverses the denominator sequence, and multiplying by z^{-N} shifts it to the right enough to make it causal. Equation 9.7 can be written for the allpass network in the form

$$H(z) = \frac{a_0 + a_1 z^{-1} + \cdots + a_{N-1} z^{-N+1} + z^{-N}}{1 + a_{N-1} z^{-1} + \cdots + a_1 z^{-N+1} + a_0 z^{-N}} \tag{9.9}$$

Here is an example to illustrate several aspects of this filter.

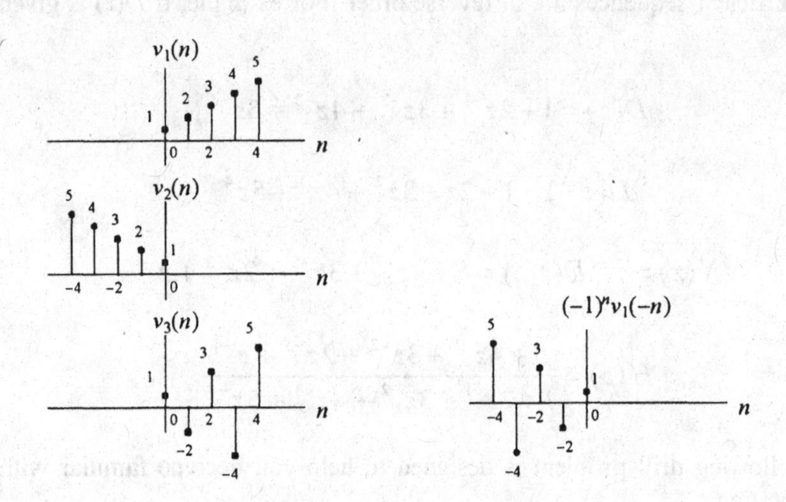

Figure 9.10. Functions for Drill 9.1.

Example 9.1. Find and plot the pole-zero plot, the rubber sheet, and the frequency response of the allpass filter given by

$$H(z) = \frac{4+6z^{-1}+4z^{-2}+z^{-3}}{1+4z^{-1}+6z^{-2}+4z^{-3}}$$

Solution: MATLAB can locate the poles and zeros by the following commands:

Poles: `roots([1 4 6 4])` gives three poles, –0.5, and $-0.5 \pm j0.5$.
Zeros: `roots([4 6 4 1])` gives three zeros, –2, and $-1 \pm j1$.

Figure 9.11 displays these poles and zeros.

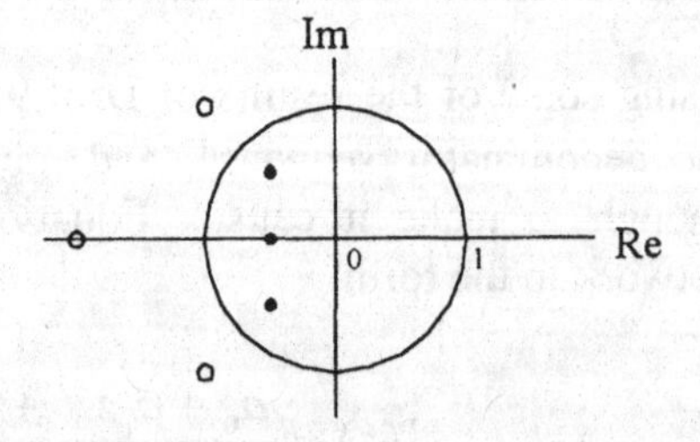

Figure 9.11. Poles and zeros.

Figure 9.12 displays the rubber sheet. The zeros at $-1 \pm j1$ and at -2 are clearly visible. The poles inside the unit circle have been suppressed to emphasize the allpass nature of this filter. The unit circle is clearly flat in this diagram, illustrating equal response at all values of Ω.

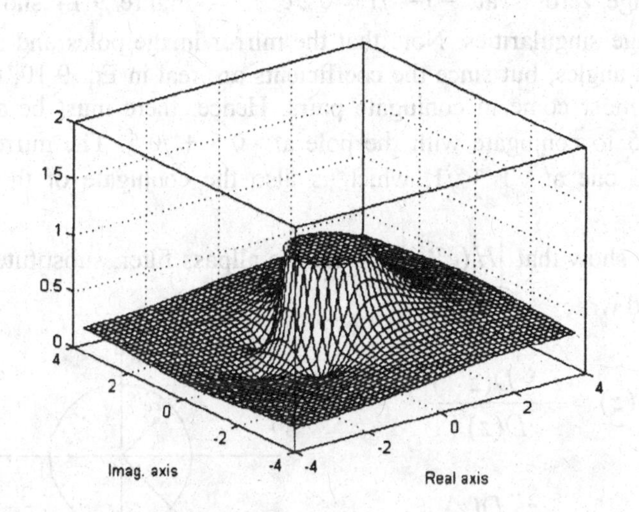

Figure 9.12. Rubber sheet.

Figure 9.13 shows the frequency response over the range $(-\pi < \Omega < \pi)$. Note that the magnitude response is constant for all values of Ω, but the phase changes from -3π to 3π. (MATLAB adjusts the angle by modulo 2π so that it ranges from $-\pi$ to π.)

■

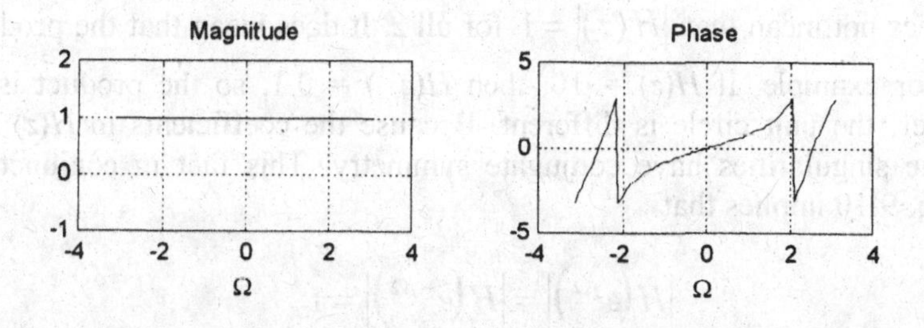

Figure 9.13. Magnitude and phase for the allpass filter.

The term *mirror image* means that if $z_p = re^{j\theta}$ is a pole of a real-coefficient allpass transfer function, it has a zero at $z_0 = \frac{1}{r}e^{-j\theta}$. Consider the pole at $-0.5 + j0.5 = \left(1/\sqrt{2}\right)e^{135^0}$ in Example 9.1. The corresponding mirror image zero is at $-1 - j1 = \sqrt{2}\,e^{-j135^0}$. Figure 9.14 shows these mirror image singularities. Note that the mirror image poles and zeros are at different angles, but since the coefficients are real in Eq. 9.10, the poles and zeros must come in conjugate pairs. Hence, there must be a pole at $-0.5 - j0.5$ to conjugate with the pole at $-0.5 + j0.5$. The mirror image zero is the one at $-1 + j1$, which is also the conjugate of the zero at $-1 - j1$.

To show that $\left|H(e^{j\Omega})\right| = 1$ for the allpass filter, substitute Eq. 9.8 into 9.9 and write

$$H(z) = \frac{z^{-N}D(z^{-1})}{D(z)}$$

Then

$$H(z^{-1}) = \frac{z^{N}D(z)}{D(z^{-1})}$$

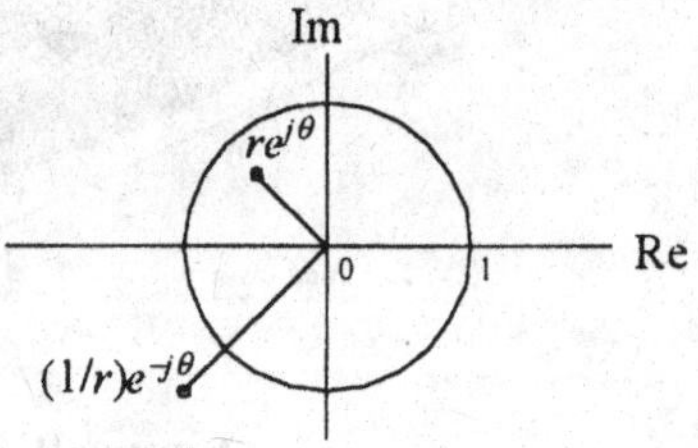

Fig. 9.11. Mirror symmetry.

Therefore,

$$H(z)H(z^{-1}) = \frac{z^{-N}D(z^{-1})}{D(z)}\,\frac{z^{N}D(z)}{D(z^{-1})} = 1 \qquad (9.10)$$

This does not mean that $\left|H(z)\right| = 1$ for all z. It does mean that the product is 1. For example, if $H(z) = 10$, then $H(z^{-1}) = 0.1$, so the product is 1. However, the unit circle is different. Because the coefficients in $H(z)$ are real, the singularities have conjugate symmetry. This fact in conjunction with Eq. 9.10 implies that

$$\left|H\left(e^{j\Omega}\right)\right| = \left|H\left(e^{-j\Omega}\right)\right| = 1$$

Hence, Eq. 9.9 represents an allpass network.

Another important concept is that of minimum-phase networks. The idea here is that the phase suffers minimum variation over the frequency interval $0 < \Omega < \pi$. To illustrate, consider the FIR filter $H_0(z) = 4 + 6z^{-1} + 4z^{-2} + z^{-3}$. This has zeros at $z = -0.5$ and $-0.5 \pm j0.5$, along with a third-order pole at the origin. Figure 9.15 shows the impulse response and the pole-zero plot.

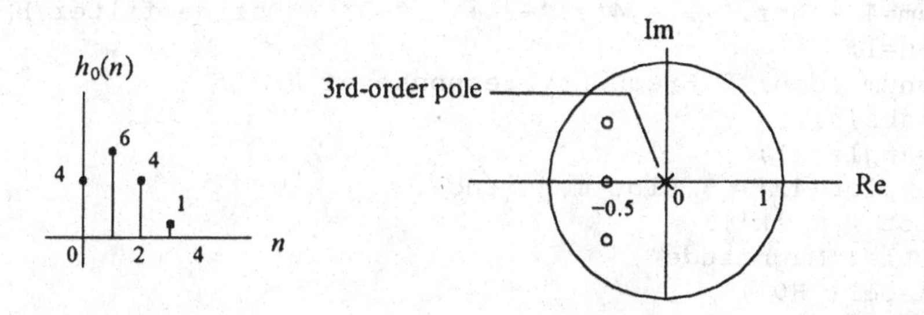

Figure 9.15. Impulse response and pole-zero plot for $H_0(z)$.

Figure 9.16 shows the pole-zero plots for two related filters, $H_1(z) = 1 + 4z^{-1} + 6z^{-2} + 4z^{-3}$ and $H_2(z) = 1 + 2.5z^{-1} + 3z^{-2} + z^{-3}$. These three filters have identical magnitude response, but different phase response, as we now show.

Example 9.2. Plot the magnitude and phase response for the filters in Figs. 9.15 and 9.16.

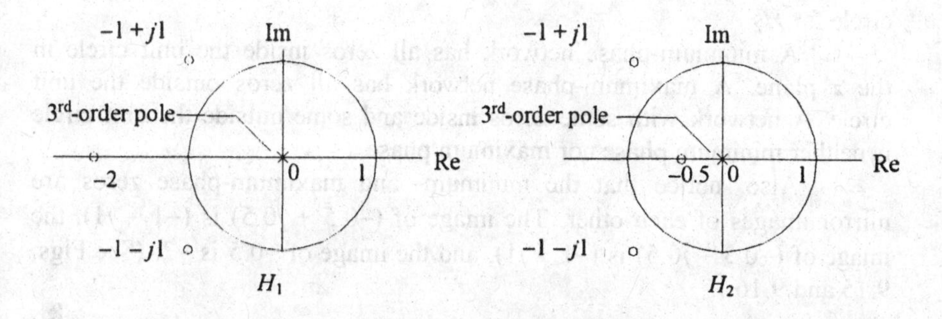

Figure 9.16. pole-zero plots for $H_1(z)$ and $H_2(z)$.

Solution: The following MATLAB program will be used in several applications. It plots the magnitude and phase for a filter specified by its transfer function. All three rows of Fig. 9.17 are produced by repeating this one section of the program.

```
w=0:0.05:pi; % Define Ω.
z=exp(j*w);  % Set z = exp(jΩ).
num=4 + 6*z.^-1 + 4*z.^-2 + z.^-3; % Define filter H0.
den=1;
h=num./den; % Frequency response of H0.
m=abs(h);
a=angle(h);
subplot(421) % Plot magnitude.
plot(w,m,'k');
title('Magnitude')
ylabel('H0')
xlabel('\Omega')
grid on
subplot(422)
plot(w,a,'k') % Plot phase.
title('Phase')
xlabel('\Omega')
grid on
```

Notice that the three magnitude responses are identical, but the three phase responses are all different. The H_0 phase varies from 0 to about -1.7 radians. The H_1 phase varies from 0 to -3π. The H_2 phase varies from 0 to -2π. These three functions represent minimum-phase, maximum-phase, and intermediate-phase variation for this one magnitude response. Notice that all zeros are inside the unit circle for H_0, all zeros are outside the unit circle for H_1, and some zeros are inside and some outside the unit circle for H_2.

A minimum-phase network has all zeros inside the unit circle in the z plane. A maximum-phase network has all zeros outside the unit circle. A network with some zeros inside and some outside the unit circle is neither minimum phase nor maximum phase.

Also, notice that the minimum- and maximum-phase zeros are mirror images of each other. The image of $(-0.5 + j0.5)$ is $(-1 - j1)$, the image of $(-0.5 - j0.5)$ is $(-1 + j1)$, and the image of -0.5 is -2. (See Figs. 9.15 and 9.16.)

■

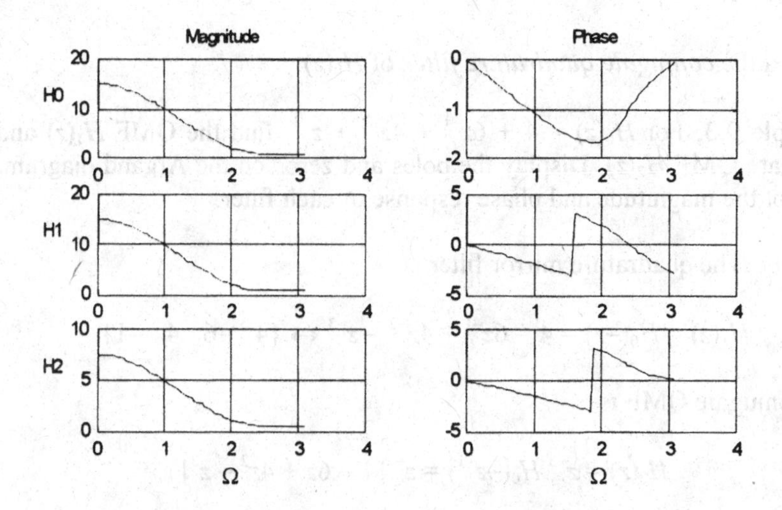

Figure 9.17. Magnitude and phase plots for H_0, H_1, and H_2.

Drill 9.2. There is one other arrangement of poles and zeros that gives the same magnitude response as $H_0(n)$ in Fig. 9.15. Find the impulse response of this filter.

Answer: $h(n) = \{1, 3, 2.5, 1\}$.

9.2. Quadrature Mirror Filters

This chapter began with an example of quadrature mirror filters (QMF). The two filters

$$H_0(z) = 1 + z^{-1} \quad \text{and} \quad H_1(z) = 1 - z^{-1}$$

form a QMF pair. Now the background is in place to understand what this means. If a filter $H_0(z)$ is FIR of order N, then

$$H_1(z) = H_0(-z)$$

is called the *quadrature mirror filter*, and

$$H_2(z) = z^{-N} H_0(-z^{-1})$$

is called the *conjugate quadrature filter* of $H_0(z)$.

Example 9.3. For $H_0(z) = 4 + 6z^{-1} + 4z^{-2} + z^{-3}$, find the QMF $H_1(z)$ and conjugate QMF $H_2(z)$. Display the poles and zeros on the Argand diagram, and plot the magnitude and phase response of each filter.

Solution: The quadrature mirror filter is

$$H_1(z) = H_0(-z) = 4 - 6z^{-1} + 4z^{-2} - z^{-3} \leftrightarrow \{4 \quad -6 \quad 4 \quad -1\}$$

The conjugate QMF is

$$H_2(z) = z^{-3}\, H_0(-z^{-1}) = z^{-3}[4 - 6z + 4z^2 - z^3]$$

$$= -1 + 4z^{-1} - 6z^{-2} + 4z^{-3} \leftrightarrow \{-1\ 4\ -6\ 4\}.$$

MATLAB finds the poles and zeros.

H_0: roots([4 6 4 1]) = -0.5, $(-0.5 \pm j0.5)$
H_1: roots([4 −6 4 −1]) = -0.5, $(-0.5 \pm j0.5)$
H_2: roots([−1 4 −6 4]) = 2, $(1 \pm j1)$

Figure 9.18 displays the poles and zeros. There is a third-order pole at the origin in each diagram. Note that the poles and zeros of H_0 and H_1 are symmetric about the origin. Also, note that the poles and zeros of H_1 and H_2 are mirror-image singularities.

The MATLAB program above plots the magnitude and phase for each filter in Fig. 9.19. Note that H_0 and H_1 show mirror image symmetry about frequency $\pi/2$ in both magnitude and phase–thus the name *quadrature mirror filters*. H_0 and H_2 have quadrature symmetry in magnitude, but not in phase. If H_0 is minimum phase, then so is H_1, and H_2 is maximum phase.

■

Example 9.4. Let $H(z) = 3 + 6z^{-1} + 2z^{-2} - z^{-3}$. Find the minimum phase filter with the same magnitude response as $H(z)$.

Solution: MATLAB finds the roots of $H(z)$, given by

roots([3 6 2 −1]) = -1.2638, -1, 0.2638.

For minimum phase, we require that all zeros must be inside (or on) the unit circle. The mirror-image zero for -1.2638 is located at $-1/1.2638 = -0.7913$. Therefore, H_0, the minimum-phase version of $H(z)$, has zeros at -0.7913, -1, and 0.2638. This gives

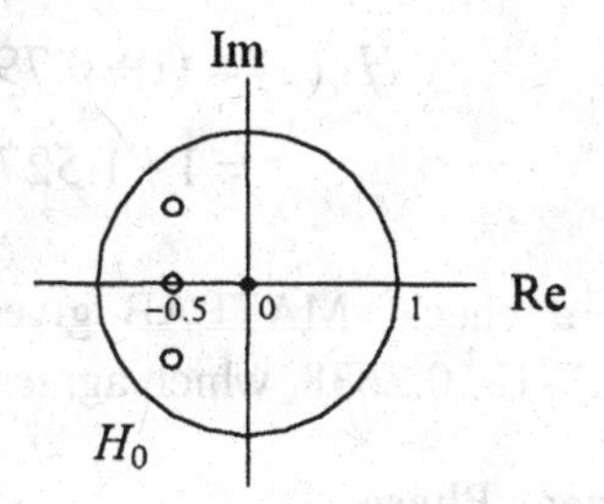

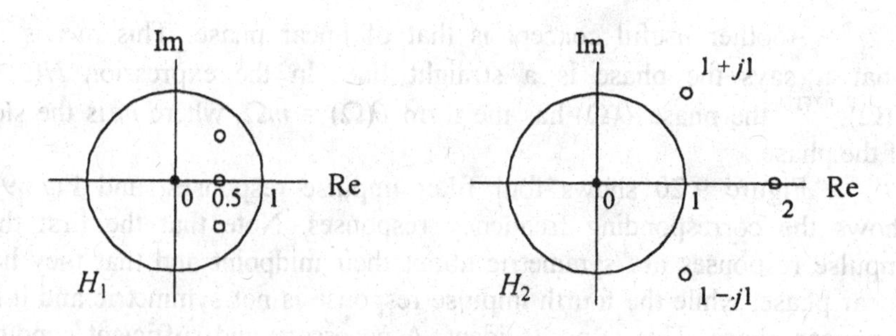

Figure 9.18. Pole-zero plot for H_0, H_1, and H_2.

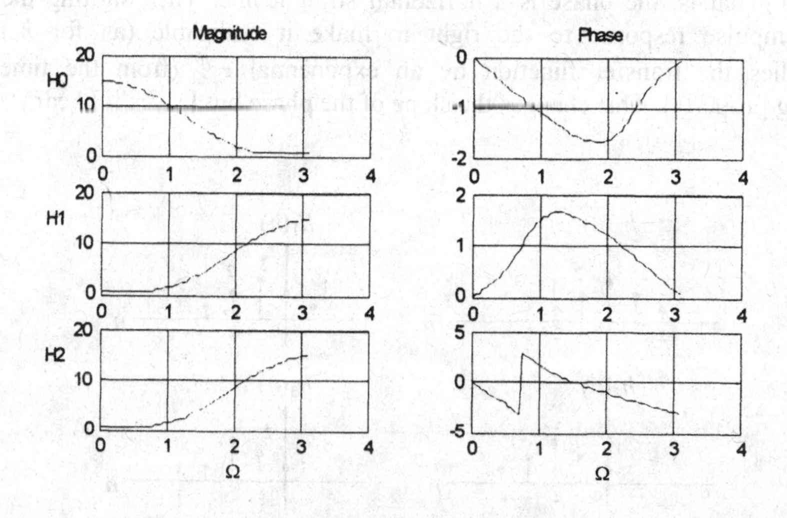

Figure 9.19. Magnitude and phase for each filter.

$$H_0(z) = (1+0.7913z^{-1})(1+z^{-1})(1-0.2638z^{-1})$$
$$= 1+1.5275z^{-1} + 0.3188z^{-2} - 0.2087z^{-3}$$

As a check, MATLAB gives roots([1 1.5275 0.3188 –0.2087]) = –1, –0.7913, 0.2638, which agrees with the specification. ■

Linear Phase

Another useful concept is that of linear phase. This means just what it says–the phase is a straight line. In the expression $H(\Omega) = A(\Omega)e^{j\theta(\Omega)}$, the phase $\theta(\Omega)$ has the form $\theta(\Omega) = m\Omega$, where m is the slope of the phase.

Figure 9.20 shows four filter impulse responses, and Fig. 9.21 shows the corresponding frequency responses. Note that the first three impulse responses are symmetric about their midpoint and that they have linear phase, while the fourth impulse response is not symmetric and it has nonlinear phase. This is no accident. A necessary and sufficient condition for linear phase is symmetry about the midpoint. (Since realizable IIR filters have no midpoint, they cannot have linear phase.) The properties of the DTFT show that a symmetrical FIR filter has linear phase. An even function of time (such as h_0) has a real transform, meaning that the phase is zero. That is, the phase is a horizontal straight line. Then shifting the even impulse response to the right to make it realizable (as for h_2) multiplies the transfer function by an exponential $e^{-j\theta}$ (from the time shifting property). This changes the slope of the phase but leaves it linear.

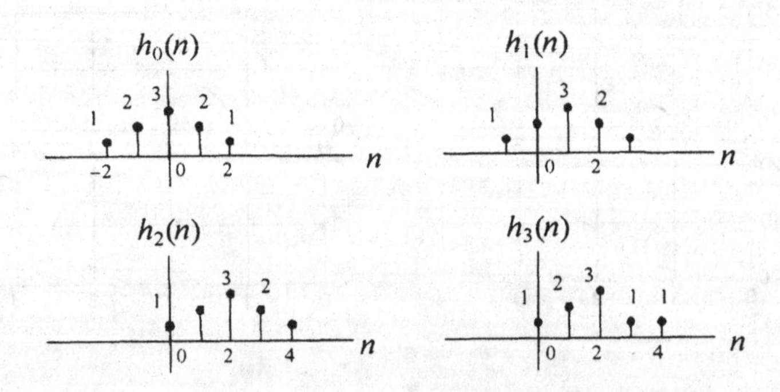

Figure 9.20. Impulse response of four filters.

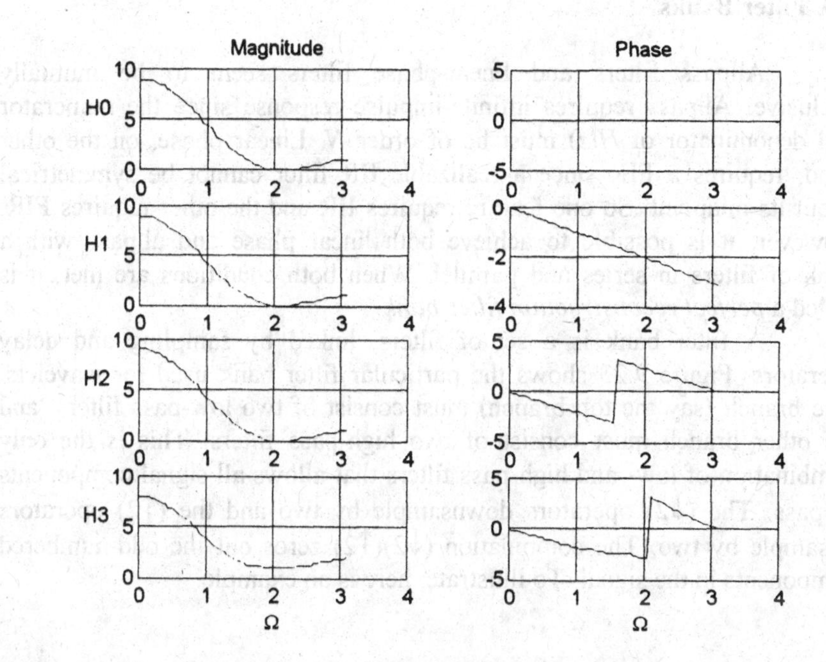

Figure 9.21. ***H_0, H_1, and H_2 are linear phase, H_3 is not.***

Drill 9.3. Plot the magnitude and phase response for the symmetric filter $h(n) = \{1\ 2\ 2\ 1\}$. What is the slope of the phase?

Answer: Fig. 9.22. The slope is -1.5Ω.

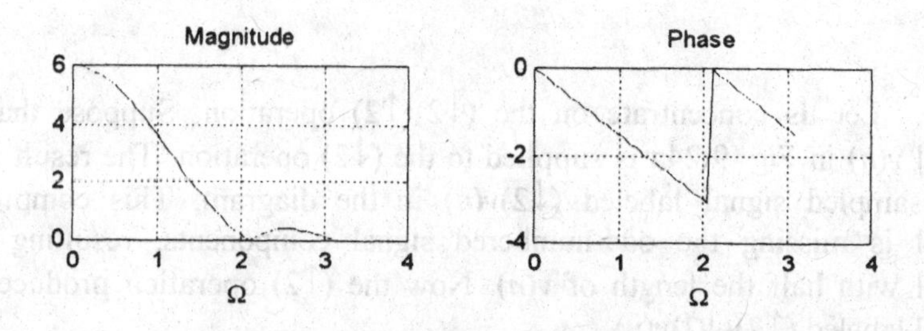

Figure 9.22. Drill 9.3 answer.

9.3. Filter Banks

Allpass filters and linear-phase filters seem to be mutually exclusive. Allpass requires infinite-impulse response since the numerator and denominator of $H(z)$ must be of order N. Linear phase, on the other hand, requires a FIR since a realizable IIR filter cannot be symmetrical about its midpoint. So one feature requires IIR and the other requires FIR. However, it is possible to achieve both linear phase and allpass with a bank of filters in series and parallel. When both conditions are met, it is called a *perfect reconstruction filter bank*.

A filter bank is a set of filters, linked by sampling and delay operators. Figure 9.23 shows the particular filter bank used for wavelets. One branch (say the top branch) must consist of two low-pass filters, and the other branch must consist of two high-pass filters. This is the only combination of low- and high-pass filters that allows all signal components to pass. The $(\downarrow 2)$ operators downsample by two and the $(\uparrow 2)$ operators upsample by two. The combination $(\downarrow 2)(\uparrow 2)$ zeros out the odd-numbered components in the signal. To illustrate, here is an example.

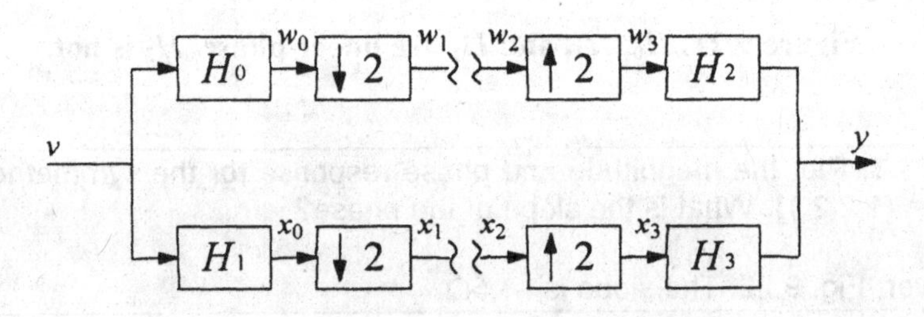

Figure 9.23. Two-channel filter bank.

Let us concentrate on the $(\downarrow 2)(\uparrow 2)$ operation. Suppose that the signal $v(n)$ in Fig. 9.24a is supplied to the $(\downarrow 2)$ operation. The result is the downsampled signal labeled $(\downarrow 2)v(n)$ in the diagram. This compressed signal is missing the odd-numbered signal components, resulting in a signal with half the length of $v(n)$. Now the $(\uparrow 2)$ operation produces the signal labeled $(\uparrow 2)(\downarrow 2)v(n)$.

Thus, the cascaded operators $(\downarrow 2)(\uparrow 2)$ retain the even-numbered components and eliminate the odd-numbered components of the signal. Figure 9.24b shows another way to arrive at this same result to help us understand why the transform of $(\uparrow 2)(\downarrow 2)v(n)$ is given by

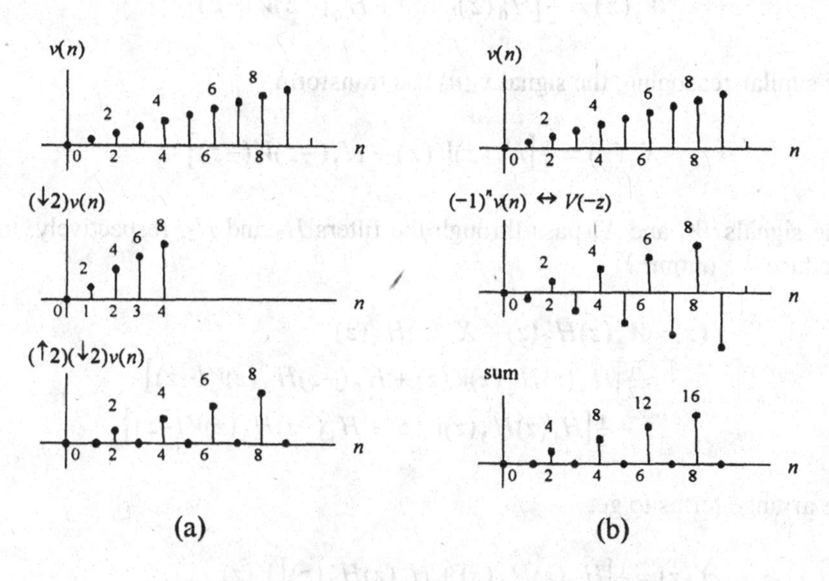

Figure 9.24. Two ways for the same result.

$$(\uparrow 2)(\downarrow 2)v(n) \leftrightarrow \tfrac{1}{2}[V(z)+V(-z)] \qquad (9.11)$$

The signals in Fig. 9.24b have the transforms

$$v(n) \leftrightarrow V(z)$$
$$(-1)^n v(n) \leftrightarrow V(-z) \text{ See Drill 9.1.}$$
$$\text{sum} \leftrightarrow V(z)+V(-z)$$

Since the sum is twice $(\uparrow 2)(\downarrow 2)v(n)$, this gives Eq. 9.11 by linearity of the transform.

Figure 9.23 has two branches. The top branch performs the operations H_0 $(\downarrow 2)(\uparrow 2)$ H_2 and the bottom branch performs the operations H_1 $(\downarrow 2)(\uparrow 2)$ H_3. For the top branch,

$$W_0(z) = H_0(z)V(z)$$

Applying the $(\downarrow 2)(\uparrow 2)$ operations gives $w_3(n)$, which according to Eq. 9.11 has the transform

$$W_3(z) = \tfrac{1}{2}\left[H_0(z)V(z) + H_0(-z)V(-z)\right]$$

By similar reasoning, the signal $x_3(n)$ has transform

$$X_3(z) = \tfrac{1}{2}\left[H_1(z)V(z) + H_1(-z)V(-z)\right]$$

The signals W_3 and X_3 pass through the filters H_2 and H_3, respectively, to produce the output Y.

$$\begin{aligned} Y(z) &= W_3(z)H_2(z) + X_3(z)H_3(z) \\ &= \tfrac{1}{2}\left[H_0(z)H_2(z)V(z) + H_0(-z)H_2(z)V(-z)\right] \\ &\quad + \tfrac{1}{2}\left[H_1(z)H_3(z)V(z) + H_1(-z)H_3(z)V(-z)\right] \end{aligned}$$

Re-arrange terms to get

$$\begin{aligned} Y(z) &= \tfrac{1}{2}\left[H_0(z)H_2(z) + H_1(z)H_3(z)\right]V(z) \\ &\quad + \tfrac{1}{2}\left[H_0(-z)H_2(z) + H_1(-z)H_3(z)\right]V(-z) \\ &= \tfrac{1}{2}F_0(z)V(z) + \tfrac{1}{2}F_1(z)V(-z) \end{aligned}$$

For perfect reconstruction with ℓ time delays, $Y(z)$ must be $z^{-\ell}V(z)$. In the absence of the ($\downarrow$2)($\uparrow$2) operations, the filter bank has the transfer function

$$F_0(z) = H_0(z)H_2(z) + H_1(z)H_3(z) \tag{9.12}$$

This is called the *distortion term*. The *alias term*

$$F_1(z) = H_0(-z)H_2(z) + H_1(-z)H_3(z) \tag{9.13}$$

is due to aliasing. For perfect reconstruction, the distortion term should represent pure delay, and the aliasing term should be zero.

$$F_0(z) = H_0(z)H_2(z) + H_1(z)H_3(z) = 2z^{-\ell} \tag{9.14}$$

$$F_1(z) = H_0(-z)H_2(z) + H_1(-z)H_3(z) = 0 \tag{9.15}$$

This gives two relationships between the four filters, leaving us free to choose two filters from other considerations. One scheme defines

$$H_2(z) = H_1(-z) \quad \text{and} \quad H_3(z) = -H_0(-z) \tag{9.16}$$

which makes the alias term zero.

Having reached this point, consider once again the distortion term.

$$H_0(z)H_2(z) + H_1(z)H_3(z)$$

The left term is the top branch, and the right term is the bottom branch in Fig. 9.23 (in the absence of the two sampling operations). Now comes a definition that allows us to simplify the distortion term.

$$P_0(z) = H_0(z)H_2(z) \qquad P_1(z) = H_1(z)H_3(z) \tag{9.17}$$

P_0 is the top branch product, or low-pass filter, and P_1 is the bottom or high-pass filter product. Figure 9.25 graphically illustrates this substitution. Substituting Eq. 9.16 into these two equations gives

$$P_0(z) = H_0(z)H_1(-z) \qquad P_1(z) = -H_0(-z)H_1(z)$$

so

$$-P_0(-z) = -H_0(-z)H_1(z) = P_1(z)$$

This simplifies the distortion term to

$$H_0(z)H_2(z) + H_1(z)H_3(z) = P_0(z) - P_0(-z) = 2z^{-\ell} \tag{9.18}$$

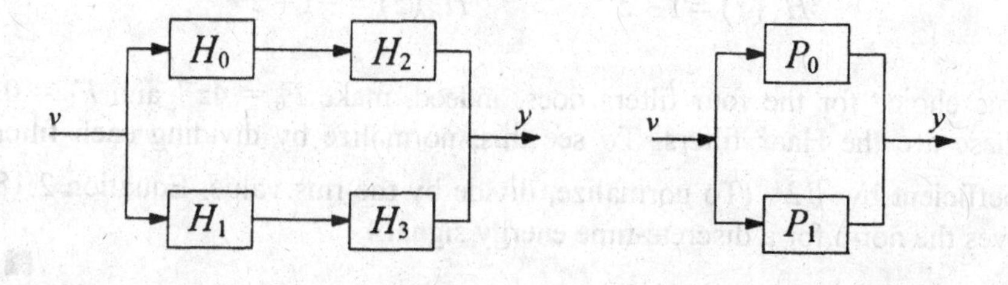

Figure 9.22. Distortion term: $P_0 = H_0H_2$ and $P_1 = H_1H_3$

The design of a perfect reconstruction filter bank reduces to two steps:

Step 1. Design a low-pass filter P_0 satisfying Eq. 9.18.
Step 2. Factor P_0 into H_0H_2. Then use the conditions in Eq. 9.16 to find H_1 and H_3.

Note that P_0 is the product of two low-pass filters, and that P_1 is the product of two high-pass filters. Thus, P_0 is itself a low-pass filter and P_1 is a high-pass filter. There are many ways to choose P_0, but one sensible way is to place one or more zeros at $z = -1$ (because we want this to be a LPF).

Example 9.5. Let $P_0(z) = (1 + z^{-1})^2$, thus having two zeros at $z = -1$. Show that Eq. 9.18 is satisfied and find all four filters.

Solution:

$$P_0(z) = 1 + 2z^{-1} + z^{-2}$$
$$P_0(-z) = 1 - 2z^{-1} + z^{-2}$$

so

$$P_0(z) - P_0(-z) = 4z^{-1}$$

The delay is correct, but the magnitude is too large. Of course, this can be corrected. Ignoring this discrepancy for the moment, factor P_0 into

$$H_0(z) = 1 + z^{-1} \qquad H_2(z) = 1 + z^{-1}$$

Equation 9.16 then gives

$$H_1(z) = 1 - z^{-1} \qquad H_3(z) = -1 + z^{-1}$$

This choice for the four filters does, indeed, make $F_0 = 4z^{-1}$ and $F_1 = 0$. These are the Haar filters. To see this, normalize by dividing each filter coefficient by $\sqrt{2}$. (To normalize, divide by the rms value. Equation 2.18 gives the norm for a discrete-time energy signal.)

■

Example 9.6. Since $(1 + z^{-1})^2$ worked well for P_0 in the preceding example, expand on this theme. Try $P_0(z) = (1 + z^{-1})^4$ and see if Eq. 9.18 is satisfied.

Solution:
$$P_0(z) = 1 + 4z^{-1} + 6z^{-2} + 4z^{-3} + z^{-4}$$
$$P_0(-z) = 1 - 4z^{-1} + 6z^{-2} - 4z^{-3} + z^{-4}$$

Therefore,

$$P_0(z) - P_0(-z) = 8z^{-1} + 8z^{-3}$$

Since this is not a single term of the form $kz^{-\ell}$, this filter bank cannot be a perfect reconstruction filter.

■

It turns out that no other filters of the form $P_0(z) = (1 + z^{-1})^N$ satisfy Eq. 9.18, so there is more to perfect reconstruction filter design than meets the eye. Once the connection between wavelets and filters was discovered, most of the work on wavelets was devoted to this problem of filter design. For each perfect reconstruction filter there is a basis for wavelets (i.e., there is a one-to-one correspondence between filters and wavelet bases). Here is one particularly good way to design filters. Let

$$P_0(z) = (1 + z^{-1})^{2p} Q(z) \tag{9.19}$$

where $Q(z)$ is the filter of order $2p - 2$ chosen to satisfy Eq. 9.18. This means that in Example 9.4 we should choose

$$Q(z) = a_0 + a_1 z^{-1} + a_2 z^{-2}$$

This makes

$$P_0(z) = \left(1 + 4z^{-1} + 6z^{-2} + 4z^{-3} + z^{-4}\right)\left(a_0 + a_1 z^{-1} + a_2 z^{-2}\right) \tag{9.20}$$

The $2p - 1$ odd powers in P_0 provide enough information to solve for the coefficients in $Q(z)$.

Example 9.7. Find the coefficients in $Q(z)$ to make

$$P_0(z) = \left(1 + z^{-1}\right)^4 Q(z)$$

a perfect reconstruction filter.

Solution: Carry out the multiplication in Eq. 9.20 to get

$$P_0(z) = a_0 + (4a_0 + a_1)z^{-1} + (6a_0 + 4a_1 + a_2)z^{-2}$$
$$+ (4a_0 + 6a_1 + 4a_2)z^{-3} + (a_0 + 4a_1 + 6a_2)z^{-4}$$
$$+ (a_1 + 4a_2)z^{-5} + a_2 z^{-6}$$

and

$$P_0(-z) = a_0 - (4a_0 + a_1)z^{-1} + (6a_0 + 4a_1 + a_2)z^{-2}$$
$$- (4a_0 + 6a_1 + 4a_2)z^{-3} + (a_0 + 4a_1 + 6a_2)z^{-4}$$
$$- (a_1 + 4a_2)z^{-5} + a_2 z^{-6}$$

Thus

$$P_0(z) - P_0(-z) = 2(4a_0 + a_1)z^{-1} + 2(4a_0 + 6a_1 + 4a_2)z^{-3}$$
$$+ 2(a_1 + 4a_2)z^{-5}$$

According to Eq. 9.18, this difference should equal $2z^{-3}$. Set this last equation equal to $2z^{-3}$ and equate like coefficients to get

$$4a_0 + a_1 = 0$$
$$4a_0 + 6a_1 + 4a_2 = 1$$
$$a_1 + 4a_2 = 0$$

This gives $a_0 = -1/16$, $a_1 = 4/16$, and $a_2 = -1/16$. Thus

$$P_0(z) = H_0(z)H_2(z) = \frac{1}{16}(1 + z^{-1})^4(-1 + 4z^{-1} - z^{-2})$$

In addition to the four zeros at $z = -1$, P_0 has zeros at $z = 2 \pm \sqrt{3}$. Choosing two of the zeros at $z = -1$ plus the minimum-phase zero at $z = 2 - \sqrt{3}$ for H_0 gives

$$H_0(z) = 0.0625 + 0.10825z^{-1} + 0.0290z^{-2} - 0.01675z^{-3}$$

This is the db(2) filter originally designed by Ingrid Daubechies.

■

Here is a program to check that a given filter h_0 satisfies Eq. 9.14. For $P_0 = H_0H_2$, assume that H_0 and H_2 are of equal length, where H_0 is minimum phase and H_2 is maximum phase. (There are other possibilities.) In other words, the coefficients in H_2 are the reverse of those in H_0. Substituting Eq. 9.16 gives the general scheme

$$\begin{aligned} h_0 &= [a_0 \quad a_1 \quad a_2 \quad a_3] \\ h_1 &= [a_3 \quad -a_2 \quad a_1 \quad -a_0] \\ h_2 &= [a_3 \quad a_2 \quad a_1 \quad a_0] \\ h_3 &= [-a_0 \quad a_1 \quad -a_2 \quad a_3] \end{aligned} \tag{9.21}$$

This generalizes to any length since the scheme is derived from Eqs. 9.14 and 9.16. As written, the following program checks a filter h_0 of length 6 [the db(3) filter]. The display follows the program. All terms in the y column are zero except for the middle term, indicating perfect reconstruction.

```
N=6;
a0=0.2352;   % define coefficients a0 through a5.
a1=0.5706;
a2=0.3252;
a3=-0.0955;
a4=-0.0604;
a5=0.0249;
h0=[a0 a1 a2 a3 a4 a5];  % define h0
h2=fliplr(h0);           % define h2
p0=conv(h0,h2);  % multiply polynomials
for k = 1:N     % Eq. 9.16 defines filters h1 and h3
  h1(k)=(-1)^(k+1)*h2(k);
  h3(k)=(-1)^k*h0(k);
end
p1=conv(h1,h3);  % multiply polynomials
y=p0+p1;  % the distortion term
disp([p0' p1' y']) % display p0, p1, y.
```

```
 0.0059 -0.0059    0
 0.0000  0.0000    0
-0.0488  0.0488    0
 0.0000  0.0000    0
 0.2930 -0.2930    0
 0.5000  0.5000 1.000
 0.2930 -0.2930    0
 0.0000  0.0000    0
```

```
-0.0488  0.0488   0
 0.0000  0.0000   0
 0.0059 -0.0059   0
```

The third column in this display is the sum $y = P_0 + P_1$. Since all terms except the middle term are zero this is a perfect reconstruction filter.

Chapter 10
Practical Wavelets and Filters

There are usually links between different paradigms in science and engineering, but the correspondence between signal theory and wavelets is astounding. This section combines the concepts of sample rate change (Chapter 6), quadrature mirror filters (Chapter 9), and wavelets (Chapter 8). This combination allows us to calculate wavelet coefficients easily from the waveform (analysis) and reconstruct the waveform from the wavelet coefficients (synthesis).

Chapter Goals: After completing this chapter, you should be able to do the following:

- Find the scaling and wavelet coefficients at level $j - 1$ from given scaling coefficients at level j (i.e., find the Haar wavelet transform from samples of a function).
- Find the matrix of transformation for a Haar wavelet transform.

10.1. Practical Wavelets

Figure 10.1 shows scaling functions and wavelets for level 0 (φ_{00} and ψ_{00}), level 1 (φ_{10}, φ_{11}, ψ_{10}, and ψ_{11}), up to level n, where there are 2^n scaling functions and 2^n wavelets. In calculating the wavelet transform, both φ_{00} and ψ_{00} are used, but from then on only wavelets are used. We can start at the bottom and work our way up, calculating first c_{00} and d_{00}, then d_{10}, and d_{11}, then d_{20}, d_{21}, and so on, until reaching level n. Alternatively, we can start at the top and calculate the level n wavelet coefficients and work our way down to the bottom of the diagram. It really doesn't matter if Eq. 8.11 does the calculation. Either way there are $2^{n+1} - 1$ wavelet coefficients and 1 scaling function coefficient for the n-level transform.

It so happens that the scaling function coefficients at level $n + 1$ can be used to calculate both the scaling function coefficients and the wavelet coefficients at level n. This uses the proper filter coefficients (those in Eqs. 8.13 and 8.14). Once this is accomplished, these same filter coefficients calculate both the scaling and wavelet coefficients at level $n - 1$ from the level-n scaling functions. This process continues down to level 0. In other words, we can calculate the entire wavelet transform for level n if we know the scaling coefficients at level $n + 1$ plus the filter coefficients. Notice what this says. Use scaling functions (not wavelets) in conjunction with filter coefficients to calculate both scaling and wavelet coefficients at the next lower level.

In going down the diagram in Fig. 10.1 from level n to level 0, the width of the scaling functions (and wavelets) doubles from one level to another. This is similar to downsampling by a factor of 2. In fact, there is more than similarity here. Downsampling and filtering with QMF filters is the usual way to calculate wavelet coefficients. Notice that there are 2^{n+1} scaling coefficients for level $n + 1$, which is the exact number of coefficients in the wavelet transform for level n. The purpose of this section is to explain how to use the scaling coefficients at level $n + 1$ to calculate the level-n transform.

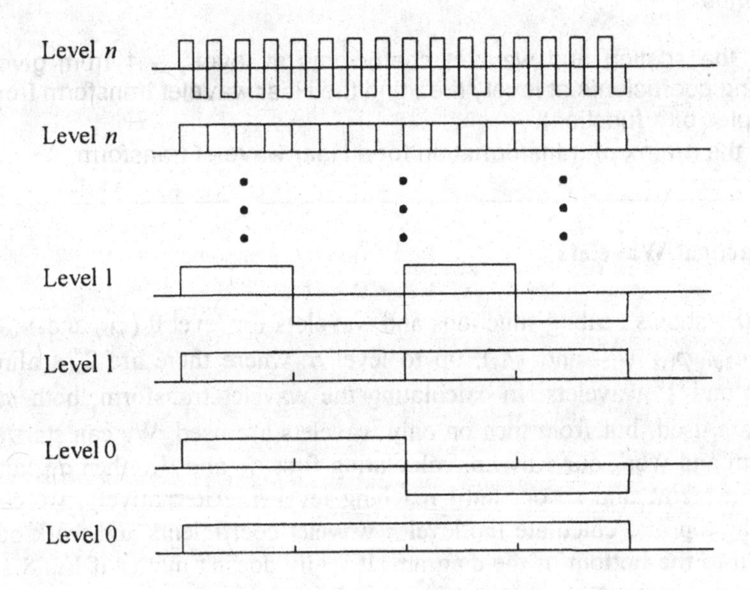

Figure 10.1. Wavelet and scaling functions, level n down to level 0.

Here is a tedious but important derivation. Recall that the forward (analysis) equations are given by

$$c_{jk} = \langle v(t) | \varphi_{jk}(t) \rangle$$
$$d_{jk} = \langle v(t) | \psi_{jk}(t) \rangle \qquad \text{Repeated (8.11)}$$

and the synthesis equation is given by

$$v(t)=\sum_{k=-\infty}^{\infty} c_{Jk}\varphi_{Jk}(t)+\sum_{j=J}^{\infty}\sum_{k=-\infty}^{\infty} d_{jk}\psi_{jk}(t) \qquad \text{Repeated (8.12)}$$

Also, Eqs. 8.13 and 8.14 provide two important relationships between scaling functions and wavelets, repeated here for convenience.

$$\varphi(t)=\sum_k h_0(k)\sqrt{2}\,\varphi(2t-k) \qquad \text{Repeated (8.13)}$$

$$\psi(t)=\sum_k h_1(k)\sqrt{2}\,\varphi(2t-k) \qquad \text{Repeated (8.14)}$$

These are called the *two-scale relation for scaling functions*, and the *two-scale relation for wavelets*, respectively. The symbols h_0 and h_1 suggest filters. Note also that these are convolution operations similar to those in Chapter 6. These equations will now be used to derive two other relationships between the wavelet coefficients given in Eq. 8.11.

Denote by V_{j+1} the set of all functions that can be represented by scaling functions at level $j+1$. That is,

$$V_{j+1}=\operatorname*{span}_k\left\{2^{(j+1)/2}\varphi(2^{j+1}t-k)\right\}$$

Meaning that V_{j+1} is a set. This set contains all the functions that can be represented by scaling functions at level $j+1$. If $v(t)\in V_{j+1}$, then $v(t)$ can be expressed with scaling functions only (no wavelets) as

$$v(t)=\sum_k c_{j+1}(k)2^{(j+1)/2}\varphi(2^{j+1}t-k)$$

But $v(t)$ can also be expressed in terms of both scaling functions and wavelets at the next lower resolution as

$$v(t)=\sum_k c_j(k)2^{j/2}\varphi(2^jt-k)+\sum_k d_j(k)2^{j/2}\psi(2^jt-k)$$

Figure 10.2 shows an example of this. The top function $v(t)$ can be represented by scaling functions φ_{3k} (not shown). As you can see, the coefficients are the eight signal values divided by $2\sqrt{2}$ (the level 3 scaling function value). $v(t)$ can also be represented by a combination of the level 2 scaling and wavelet functions shown. This also gives us eight coefficients, which represent the function in terms of the φ_{2k} and ψ_{2k} functions.

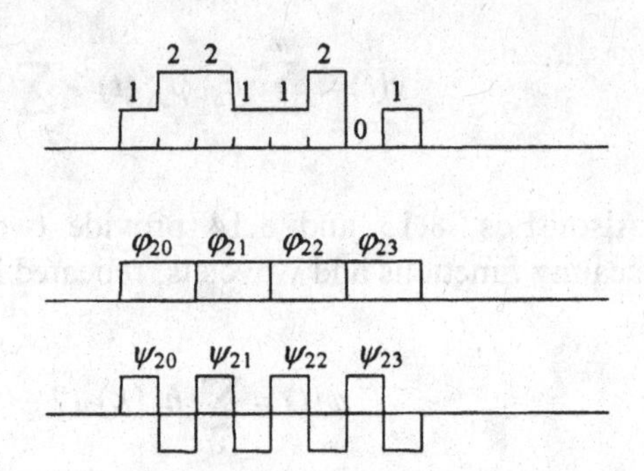

Figure 10.2. Representing $v(t)$.

Drill 10.1. Find the four scaling coefficients for φ_{2k} and four wavelet coefficients for ψ_{2k} that represent the function in Fig. 10.2. The height of each scaling function and wavelet is 2.

Answer: $c_2(k) = \{0.75, 0.75, 0.75, 0.25\}$,
$d_2(k) = \{-0.25, 0.25, -0.25, -0.25\}$.

Notice that knowing only the scaling function coefficients at level 3 provides enough information to calculate both the scaling and wavelet coefficients at level 2. This implies a direct relationship between the scaling functions at one level and the scaling functions and wavelets at another level. It is this relationship, derived below, that provides the magic in wavelet theory. Note the change in notation, from c_{jk} to $c_j(k)$, so that these equations will look more like filtering operations.

Here is the derivation: Equation 8.11 gives the c_{jk} coefficients in terms of the inner product

$$c_j(k) = \langle v(t) | \varphi_{jk}(t) \rangle = \int v(t) 2^{j/2} \varphi(2^j t - k)\, dt \qquad (10.1)$$

To find an expression for $\varphi(2^j t - k)$, start with Eq. 8.13. Scale and translate the scaling function to obtain

$$\varphi(2^j t - k) = \sum_n h_0(n) \sqrt{2} \varphi(2(2^j t - k) - n) = \sum_n h_0(n) \sqrt{2} \varphi(2^{j+1} t - 2k - n)$$

Perform a change of variable and let $m = 2k + n$; then $n = m - 2k$, giving

$$\varphi(2^j t - k) = \sum_m h_0(m - 2k)\sqrt{2}\,\varphi(2^{j+1}t - m)$$

Substitute this into Eq. 10.1 and interchange the sum and integral to get

$$c_j(k) = \sum_m h_0(m - 2k)\int v(t)2^{(j+1)/2}\varphi(2^{j+1}t - m)\,dt$$

The integral in this expression is the inner product at the scale $j + 1$ in Eq. 8.11, giving

$$c_j(k) = \sum_m h_0(m - 2k)c_{j+1}(m) \qquad (10.2)$$

A similar derivation gives the wavelet coefficient relation as

$$d_j(k) = \sum_m h_1(m - 2k)c_{j+1}(m) \qquad (10.3)$$

Equations 10.2 and 10.3 state that $c_{j+1}(k)$ provides enough information to find all lower-scale coefficients. These equations are the key to easy calculation of wavelet coefficients. From the scaling coefficients $c_{j+1}(k)$ at level $j+1$, calculate both the scaling and wavelet coefficients at level j. Calculate all lower-resolution coefficients by iteration. (Of course, the coefficients for filters h_0 and h_1 must also be known.) If we are given or can somehow find the coefficients at some level, we can find all lower-level coefficients. Here is an example.

Example 10.1. Derive the c_1, d_1, c_0, and d_0 sequences if c_2, h_0, and h_1 are given by

$$c_2 = [1 \quad 2 \quad 3 \quad 4] \quad h_0 = [1/\sqrt{2} \quad 1/\sqrt{2}] \quad h_1 = [1/\sqrt{2} \quad -1/\sqrt{2}]$$

Solution: Let us first derive the c_1 sequence from Eq. 10.2. Figure 10.3 shows $c_2(m)$, $h_0(m)$, and $h_0(m - 2)$. Setting $k = 0$ in Eq. 10.2, multiplying, and then summing gives

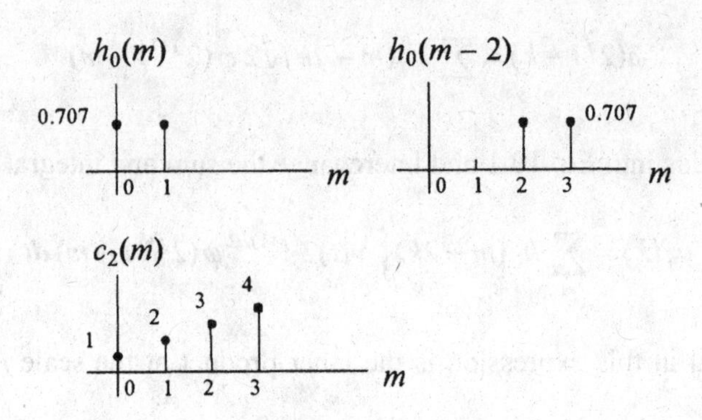

Figure 10.3. Filter and starting sequence.

$$c_1(0) = \sum_m h_0(m)c_2(m) = \frac{1}{\sqrt{2}}[1+2] = \frac{3}{\sqrt{2}}$$

Next, with $k = 1$,

$$c_1(1) = \sum_m h_0(m-2)c_2(m) = \frac{1}{\sqrt{2}}[3+4] = \frac{7}{\sqrt{2}}$$

In a similar manner, the d_1 sequence from Eq. 10.3 is given by

$$d_1(0) = \sum_m h_1(m)c_2(m) = \frac{1}{\sqrt{2}}[1-2] = -\frac{1}{\sqrt{2}}$$

$$d_1(1) = \sum_m h_1(m-2)c_2(m) = \frac{1}{\sqrt{2}}[3-4] = -\frac{1}{\sqrt{2}}$$

Finally,

$$c_0(0) = \sum_m h_0(m)c_1(m) = \frac{1}{\sqrt{2}}\left[\frac{3}{\sqrt{2}} + \frac{7}{\sqrt{2}}\right] = 5$$

$$d_0(0) = \sum_m h_1(m)c_1(m) = \frac{1}{\sqrt{2}}\left[\frac{3}{\sqrt{2}} - \frac{7}{\sqrt{2}}\right] = -2$$

■

Drill 10.2. Use the h_0 and h_1 filters in Example 10.1 and derive the c_1, d_1, c_0, and d_0 sequences if c_2 is given by

$$c_2 = [0.5 \quad 0.2 \quad -0.1 \quad 0.3]$$

Answer: 0.495, 0.1414, 0.2121, −0.2828, 0.45, 0.25.

Notice in this example that the starting sequence $c_2(m)$ has length 4. After correlation and downsampling, the resulting sequence $c_1(m)$ has length 2, and one more operation reduces the sequence to length 1. Also, notice that the process of deriving c_j from c_{j+1} is equivalent to filtering and downsampling. Correlation with $h(m)$ is convolution with $h(-m)$.

10.2. The Magic Part

Here is the magic part. Knowing the scaling coefficients c_{jk} at level j allows us to calculate both the scaling and wavelet coefficients at level $j-1$. The problem is getting started–how to find the coefficients at level j? The answer is that it is easy, and that is the magic part. Equation 8.11 says that the coefficients c_{jk} are found by the inner product:

$$c_{jk} = \langle v(t) | \varphi_{jk}(t) \rangle$$

For the Haar transform the inner product looks like Fig. 10.4. The scaling function $\varphi_{jk}(t)$ is a square wave of fixed height. When multiplied by the waveform $v(t)$ and integrated over its width, the resulting number is proportional (approximately) to the sample value of the waveform at time t_k. Therefore, the samples themselves, with appropriate modification, become the coefficients c_{jk}. The appropriate modification depends on the width of the scaling function, which depends on the sampling rate.

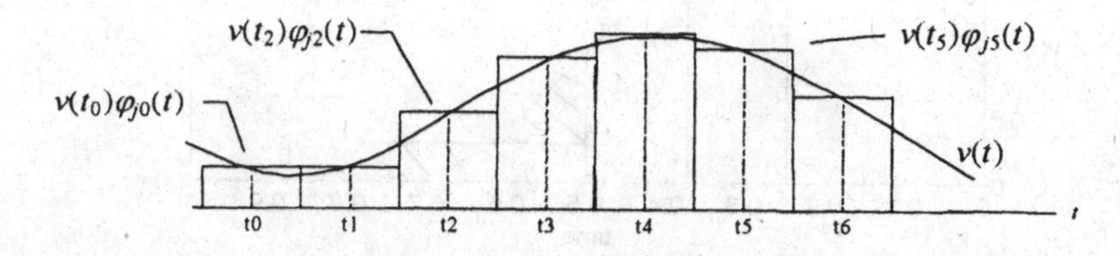

Figure 10.4. Finding the inner products c_{jk}.

The spacing between samples has everything to do with the accuracy of the wavelet representation. For a given set of coefficients c_{jk}, the best one can do is construct the staircase approximation to $v(t)$. This is Fig. 10.4 without the smooth curve $v(t)$. This agrees with intuition. The more wavelet coefficients, the better the approximation. In the limit, as the number of samples increases, the closer is the approximation.

Example 10.2. Here is an example to illustrate the discussion above and to demonstrate the kind of error obtained from using samples as the level j coefficients. Compare the coefficient values at level $j = 2$ to the modified sample values for the waveform $v(t) = 1 + \cos(2\pi t)$ over the interval $0 < t < 1$.

Solution: Figure 10.5 shows $v(t)$ with nine samples. These nine samples are located at $t = 0$, 1/8, 1/4, 3/8, 1/2, 5/8, 3/4, 7/8, and 1. Select the samples at $t = 1/8$, 3/8, 5/8, and 7/8, and delete the others to get the required samples. MATLAB does this with the statements

```
N=8;
t=linspace(0,1,N+1); %Locates 9 samples in (0,1)
v=1 + sin(2*pi*t); %produces samples of v

a=v(2:N+1);    %Eliminates first sample
a=reshape(a,2,N/2); %desired samples on top row
c=a(1,:);   %selects top row (the desired samples)
```

This gives the sample values at $t = 1/8$, 3/8, 5/8, and 7/8 as [1.7071 1.7071 0.2929 0.2929].

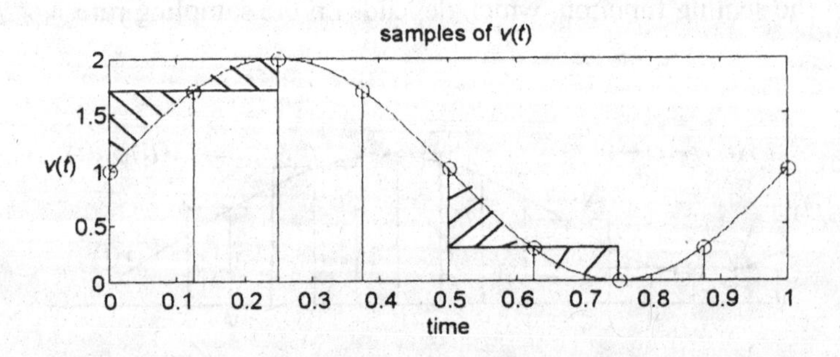

Figure 10.5. Samples of $v(t)$.

Also shown in Fig. 10.5 are the scaling functions $\varphi_{20}(t)$ and $\varphi_{22}(t)$ as horizontal lines. The shaded portions present a picture of the error introduced by using samples as the coefficients c_{20} and c_{22}. For the samples to represent the coefficients accurately, the shaded areas to the left and right of the sample should be equal. As you can see, this occurs when the function is a straight line through the sample point. There will be considerable error in this example. To obtain the coefficients, multiply the sample values by the normalizing constant 2 and multiply by the pulse width of ¼. (The normalizing constant is 2 because we have multiplied by $\sqrt{2}$ twice.) This gives [0.8536 0.8536 0.1464 0.1464] for the coefficients. These are not very accurate, but there are only four samples. The accuracy increases with the number of samples.

■

Example 10.3. Find the Haar wavelet transform for the signal in Fig. 10.6.

Solution: This example combines sample rate change with QMF filters to find wavelet coefficients. Define c_3 in MATLAB as

```
c3 = [1 2 1 -1 1 2 -1 -2]
```

This produces the signal in Fig. 10.6. Think of these as coefficients that are derived from samples of a continuous-time signal after proper modification. Also think of these as values of a discrete-time signal. In any event, label these c_{3k}, or $c_3(k)$, $k = 0, 1, \ldots, 7$.

Use the Haar wavelets and define the QMF filter bank h_0, h_1, h_2, and h_3 with the statements

```
h0=[1 1]/sqrt(2);
h1=[1 -1]/sqrt(2);
h2=h0;
h3=fliplr(h1);
```

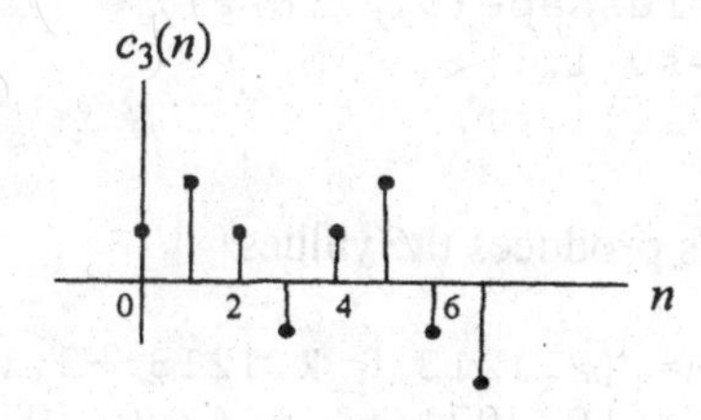

Figure 10.6. Arbitrary signal.

Figure 10.7 shows the process for finding the transform. Calculate c_2 and d_2 coefficients by filtering and downsampling. First, filter by

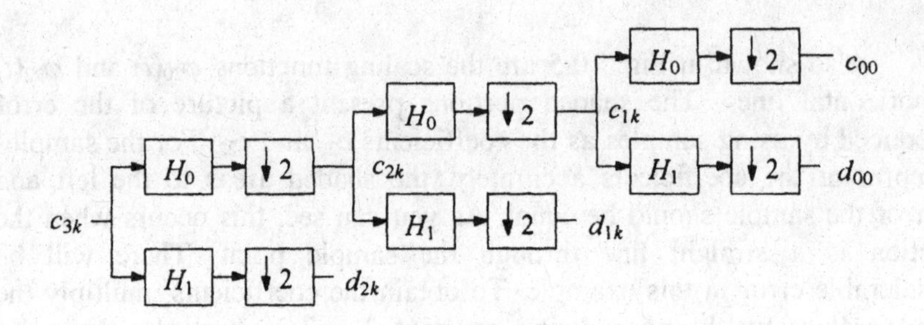

Figure 10.7. Analysis process.

```
s0=conv(h0,c3);
s1=conv(h1,c3);
```

This produces nine values for s_0 and s_1. Discard the first and last values because they represent transients. (To see why, think about signals entering and leaving a shift register.) Eliminate the first value with the statements

```
N=8;
s0=s0(:,2:N+1);
s1=s1(:,2:N+1);
```

and then eliminate the last value by downsampling.

```
s0=reshape(s0,2,N/2);
s1=reshape(s1,2,N/2);
c2=s0(1,:);
d2=s1(1,:);
```

This produces the values

```
c2 = [2.1213 0 2.1213 -2.1213]
d2 = [0.7071 -1.4142 0.7071 -0.7071]
```

This completes the first stage in Fig. 10.7. Here is the program complete to this stage.

```
c3 = [1 2 1 -1 1 2 -1 -2];
h0=[1 1]/sqrt(2);
h1=[1 -1]/sqrt(2);
h2=h0;
```

```
h3=fliplr(h1);
N=8;
s0=conv(h0,c3);     % ← filter
s1=conv(h1,c3);     % ←
s0=s0(:,2:N+1);     % ← Eliminate first value
s1=s1(:,2:N+1);     % ←
s0=reshape(s0,2,N/2);    % ← downsample
s1=reshape(s1,2,N/2);    % ←
c2=s0(1,:);         % ← gives c2
d2=s1(1,:);         % ← and d2
```

The second stage in Fig. 10.7 is identical to the first stage. Iterate by copying the last program block and making necessary changes to obtain

```
N=N/2;
s0=conv(h0,c2);     % ← filter
s1=conv(h1,c2);     % ←
s0=s0(:,2:N+1);     % ← Eliminate first value
s1=s1(:,2:N+1);     % ←
s0=reshape(s0,2,N/2);    % ← downsample
s1=reshape(s1,2,N/2);    % ←
c1=s0(1,:);         % ← gives c1
d1=s1(1,:);         % ← and d1
```

The only changes necessary were in the first three statements and in the last two statements. This produces

```
c1 = [1.5 0]
d1 = [-1.5 -3]
```

One more iteration produces

$c_{00} = 1.0607$
$d_{00} = -1.0607$

Out of all these coefficients, select the following eight to represent the original waveform c_3. The Haar wavelet transform of c_3 is given by $[c_{00}\ d_{00}\ d_{10}\ d_{11}\ d_{20}\ d_{21}\ d_{22}\ d_{23}]$ or

$$[1.0607\quad -1.0607\quad -1.5\quad -3\ 0.7071\quad -1.4142\quad 0.7071\quad -0.7071]$$

■

This completes the analysis example. Let us check that this agrees with the wavelet transform. Think of the values c_{3k} as samples of a continuous-time waveform, say $v(t)$. Figure 10.8 illustrates this concept. To calculate the wavelet coefficient c_{00}, find the inner product between $v(t)$ and $\varphi_{00}(t)$. That is, multiply this waveform by a constant and integrate over the interval. Approximate this inner product by summing the values c_{3k}, for $k = 0, 1, \ldots, 7$. This gives Sum = 3, which is not c_{00}. Divide Sum by $2\sqrt{2}$ to get $c_{00} = 1.0607$, the correct answer. The $2\sqrt{2}$ value represents the normalizing factor between φ_{00} and φ_{3k}.

The coefficient d_{00} is the inner product between $\psi_{00}(t)$ and $v(t)$. To approximate this, subtract the last four values of c_{3k} from the first four values (because $\psi_{00}(t)$ is +1 for half the period and −1 for the other half.) After division by $2\sqrt{2}$ this gives $d_{00} = -1.0607$, the correct answer. In addition, d_{10} can be calculated by the equation

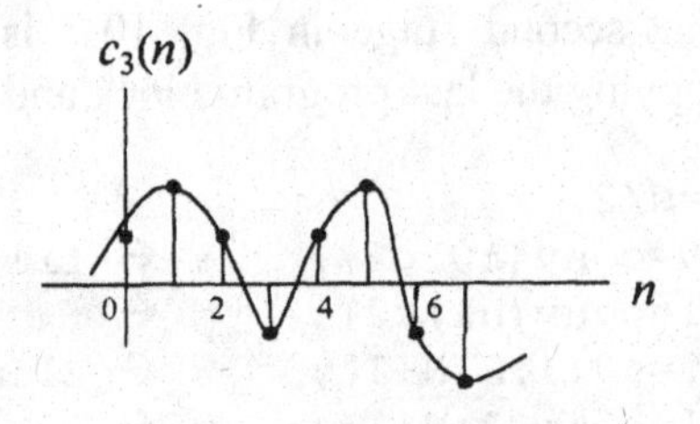

Figure 10.8. Values of c_3 as samples.

$$d_{10} = \frac{c_{30} + c_{31} - c_{32} - c_{33}}{2} = -1.5$$

with a similar formula for d_{11}. The d_{2k} coefficients are found by taking differences between successive values of c_{3k} and dividing by $\sqrt{2}$. Thus, our filtering example agrees with our previous method of calculating the wavelet coefficients.

Example 10.4. Reconstruct the waveform $c_3(n)$ in Fig. 10.6 from the wavelet coefficients.

Solution: Supply the eight coefficients $[c_{00}\ d_{00}\ d_{10}\ d_{11}\ d_{20}\ d_{21}\ d_{22}\ d_{23}]$ = [1.0607 −1.0607 −1.5 −3 0.7071 −1.4142 0.7071 −0.7071] to the filter-upsample process in Fig. 10.9. Start with c_{00} and d_{00}, upsample these to produce the signals [1.0607 0] and [−1.0607 0], and convolve these with the filters h_2 and h_3. Then sum these signals to produce c_{1k}. The convolution of $[1\ \ 1]/\sqrt{2}$ with [1.0607 0] produces [0.7500 0.7500 0].

The convolution of $[-1\ \ 1]/\sqrt{2}$ with $[-1.0607\ \ 0]$ produces $[0.7500\ \ -0.7500\ \ 0]$. Adding gives

$$[0.7500\ \ 0.7500\ \ 0] + [-0.7500\ \ -0.7500\ \ 0]. = [1.500\ \ 0\ \ 0]$$

Delete the last zero to obtain $c_{1k} = [1.500\ \ 0]$.

Here is the MATLAB program to accomplish this.

```
%Inverse process
p0=[c0 0]; %Upsample c0
p1=[d0 0]; %Upsample d0
q0=conv(p0,h2) %Convolve with h2
q1=conv(p1,h3) %Convolve with h3
c1=q0+q1;     %Add
c1=c1(:,1:N)   %Delete last zero
```

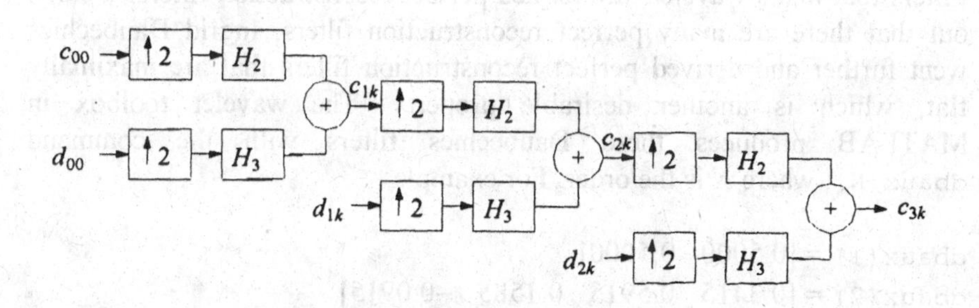

Figure 10.9. Synthesis process.

At the next stage in Fig. 10.9, the input to the $(\uparrow 2)H_2$ branch is c_{1k}. Upsample c_{1k} and convolve with $h_2(k)$. Then upsample d_{1k} and convolve with $h_3(k)$. Here is the next program segment (almost identical to the above) to do this and add the result to get c_{2k}.

```
N=2*N;
p0=[c1
  0 0];
p0=p0(:)'; %Upsample c1
p1=[d1
  0 0];
p1=p1(:)'; %Upsample d1
q0=conv(p0,h2) %Convolve with h2
q1=conv(p1,h3) %Convolve with h3
```

```
c2=q0+q1;    %Add
c2=c2(:,1:N)  %Delete last zero
```

One more iteration produces the original signal c_3. This completes the example.

■

10.3. Other Wavelets

Figure 10.7 represents the analysis process (the forward wavelet transform) as a sequence of filter-downsample operations. Figure 10.9 represents the synthesis process (the inverse wavelet transform) as a sequence of upsample-filter operations. If these two operations are placed in cascade, the output should equal the input. In fact, Examples 10.3 and 10.4 demonstrate that this is true for the Haar wavelet system. Now the question is: What other wavelets give us this desirable property?

Actually, we seldom look for wavelets, but rather look for filters. Filters that make wavelets are termed perfect reconstruction filters. It turns out that there are many perfect reconstruction filters. Ingrid Daubechies went further and derived perfect reconstruction filters that are maximally flat, which is another desirable property. The wavelet toolbox in MATLAB produces these Daubechies filters with the command `dbaux(N)`, where N is the order. For example,

`dbaux(1)` = [0.5000 0.5000]
`dbaux(2)` = [0.3415 0.5915 0.1585 −0.0915]
`dbaux(3)` = [0.2352 0.5706 0.3252 −0.0955 −0.0604 0.0249]

These coefficients all have norm equal to 0.7071, rather than 1. Notice that the Haar filter is the first Daubechies filter. Also, notice that the Nth-order filter has length $2N$. Figure 10.10 shows the Daubechies scaling function and wavelet of order 2. These two diagrams were generated by upsampling and convolving as follows: To generate the scaling function:

1. Let h = dbaux(2).
2. Let a = fliplr(h).
3. Upsample $a(n)$ by 2.
4. Convolve h with a. i.e., let a = conv(h,a).
5. Repeat steps 3 and 4 several times.

To generate the wavelet, multiply $a(n)$ by $(-1)^n$ in step 2. Notice that the length of a doubles each time, so this is exponential growth. Steps

3 and 4 cannot be repeated too many times. Here is the program that generated Fig. 10.10:

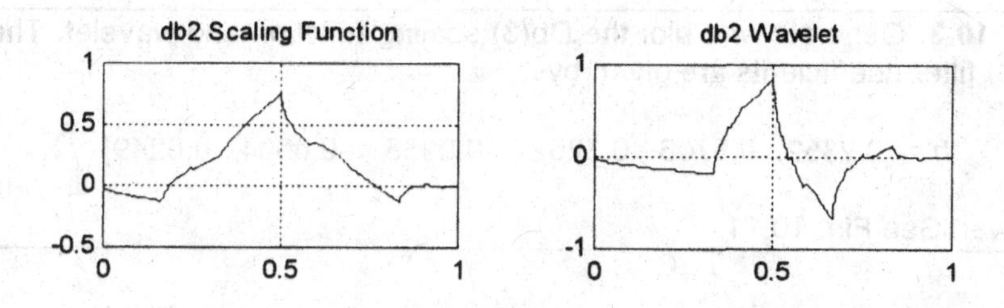

Figure 10.10. db2 functions.

```
k=2; % normalizing constant
h=[0.3415 0.5915 0.1585 -0.0915]; %filter h
a=fliplr(h); % initial scaling function
w=[-0.0915 -0.1585 0.5915 -0.3415]; %initial wavelet
for m = 1:7 %iterate
 b=[  a
  zeros(size(a))]; % upsample a
 a=b(:)'; %insert alternating zeros
 a=conv(h,a); %convolve h with a
 a=k*a;
 n=length(a);
 a=a(1,1:n-1); %delete zero term
end
n=length(a); %plot scaling function
x=linspace(0,1,n);
subplot(321)
plot(x,a,'k')
grid on
title('db2 scaling functin')
for m=1:7 %calculate wavelet w
 b=[ w
  zeros(size(w))];
 w=b(:)'; %insert alternating zeros
 w=conv(h,w); %convolve h with w
 w=k*w;
 n=length(w);
 w=w(1,1:n-1);
end
n=length(w); %plot wavelet
x=linspace(0,1,n);
```

```
subplot(322)
plot(x,w,'k')
grid on;  title('db2 wavelet')
```

Drill 10.3. Generate and plot the Db(3) scaling function and wavelet. The Db(3) filter coefficients are given by

$$h = [0.2352 \quad 0.5706 \quad 0.3252 \quad -0.0955 \quad -0.0604 \quad 0.0249]$$

Answer: See Fig. 10.11.

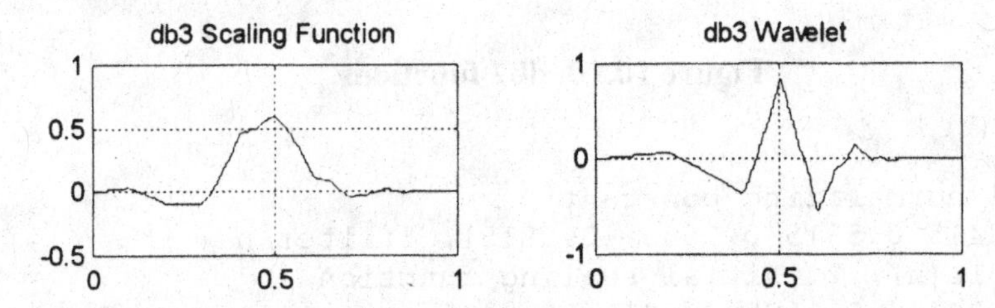

Figure 10.11. db3 functions

Biorthogonality

An orthogonal multiresolution analysis (MRA) defines W_j as the orthogonal complement of V_j in V_{j+1}. A sufficient condition for an MRA to be orthogonal is

$$W_0 \perp V_0$$

The orthogonality property puts a strong limitation on the construction of wavelets. The Haar wavelet is the only real-valued wavelet with finite extent that is symmetric and orthogonal. The generalization to biorthogonal wavelets gains more flexibility.

Our purpose in this section is to provide an example of this. Our purpose in Section 3.4 was to provide background for this discussion. Recall that $\{x_1, x_2, \ldots, x_N\}$ and $\{y_1, y_2, \ldots, y_N\}$ are said to be reciprocal sets of vectors if they satisfy

$$\langle x_i | y_j \rangle = \delta_{ij}, \quad i, j, = 1, 2, \cdots, N \qquad \text{Repeated (3.5)}$$

When applied to wavelets, this *biorthogonality condition* means that a dual scaling function $\tilde{\varphi}$ and dual wavelet $\tilde{\psi}$ exist that generate a dual MRA with subspaces $\tilde{V}_j$ and $\tilde{W}_j$ such that

$$\tilde{V}_j \perp W_j \quad \text{and} \quad V_j \perp \tilde{W}_j \tag{10.4}$$

To give some inkling of what is involved here, consider the set of all discrete-time signals of length 8. Figure 10.12 shows one basis for V_3 consisting of a basis for V_2 on the left plus the basis for W_2 on the right. This is the usual basis for Haar wavelets.

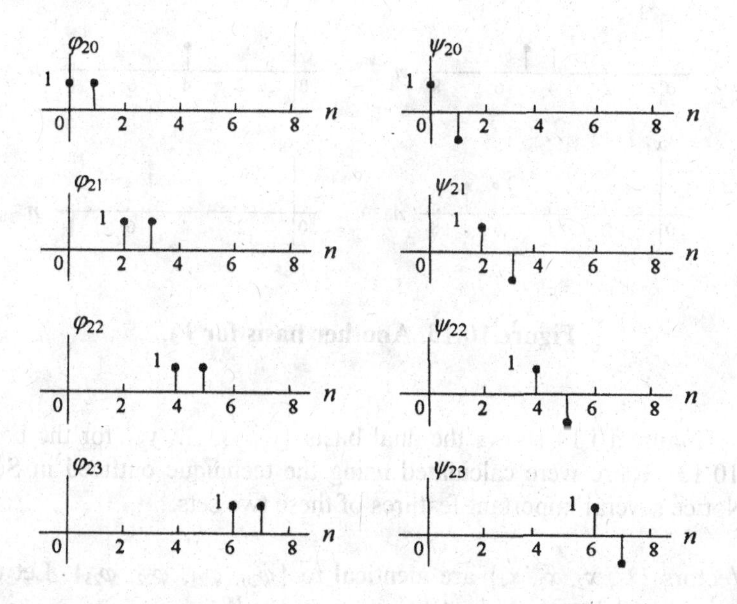

Figure 10.12. Usual basis for V_3.

The basis vectors in Fig. 10.12 are mutually orthogonal. (They would be orthonormal if scaled properly.) Except for scale, the reciprocal basis vectors are the same. Contrast this with the basis set in Fig. 10.13. Here, all eight basis vectors are not mutually orthogonal. They are independent, so they form a basis, but they are not all orthogonal to each other. For example, x_1 is not orthogonal to x_5. This means that the reciprocal basis vectors will not be the same as the x_i's.

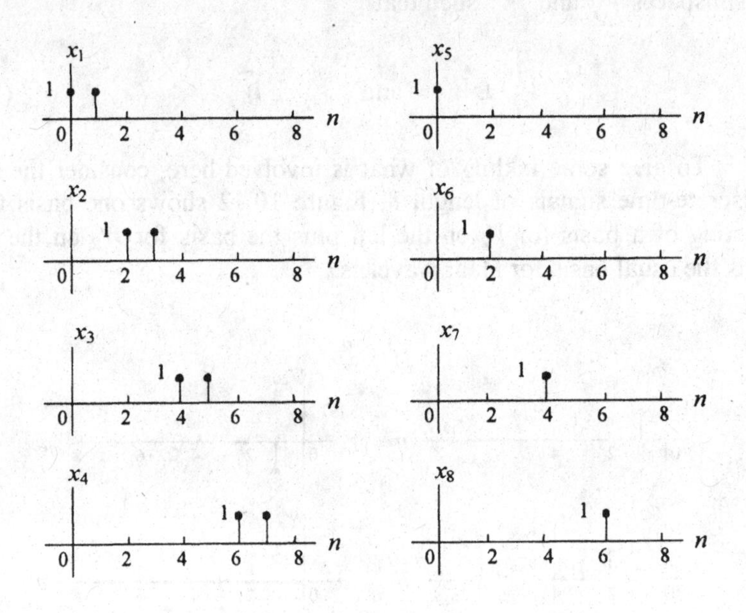

Figure 10.13. Another basis for V_3.

Figure 10.14 shows the dual basis $\{y_1, y_2, \ldots, y_8\}$ for the basis in Fig. 10.13. These were calculated using the technique outlined in Section 3.4. Notice several important features of these two sets.

1. Vectors $\{x_1, x_2, x_3, x_4\}$ are identical to $\{\varphi_{20}, \varphi_{21}, \varphi_{22}, \varphi_{23}\}$. Let us set $\{x_1, x_2, x_3, x_4\} = V_2$, and set $\{x_5, x_6, x_7, x_8\} = W_2$.
2. Vectors $\{y_5, y_6, y_7, y_8\}$ are identical to $\{\psi_{20}\ \psi_{21}, \psi_{22}, \psi_{23}\}$. Let us set $\{y_1, y_2, y_3, y_4\} = \widetilde{V}_2$, and set $\{y_5, y_6, y_7, y_8\} = \widetilde{W}_2$.

Then Eq. 10.4 is satisfied, that is,

$$\widetilde{V}_2 \perp W_2 \quad \text{and} \quad V_2 \perp \widetilde{W}_2$$

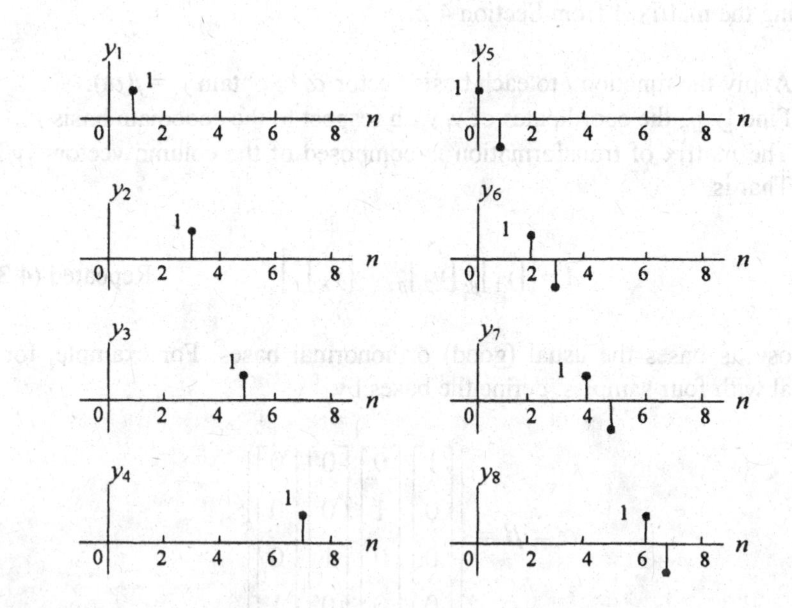

Figure 10.14. Reciprocal basis for Fig. 10.13.

Note that $V_3 = V_2 + W_2$, although the sum is not orthogonal. Comparing these vectors to those in Fig. 10.13 shows that $\psi_{20} = 2x_5 - x_1$, and so on. Similarly, $\varphi_{20} = 2y_1 + y_5$, etc. Compute angles using Eq. 2.22. For example, the angle between x_1 and x_5 is given by

$$\cos\theta = \frac{\langle x_1 | x_5 \rangle}{\|x_1\| \|x_5\|} = \frac{1}{2}$$

This gives $\theta = \pm 60^\circ$.

10.4. Matrix of Transformation

The Haar wavelet transform is a linear transformation between two finite-dimensional vector spaces. Therefore, it can be represented by a matrix. This section concentrates on this view.

Call the matrix of transformation A. Recall from Section 4.2 that this matrix depends only on the transformation f and on the bases α and β

for the domain and codomain, respectively. Here is the algorithm for finding the matrix A from Section 4.2.

1. Apply the function f to each basis vector α_i to obtain $y_i = f(\alpha_i)$.
2. Find $[y_i]_\beta$, the coordinates of y_i with respect to the codomain basis β.
3. The matrix of transformation is composed of the column vectors $[y_i]_\beta$. That is,

$$A = \left[[y_1]_\beta \ [y_2]_\beta \ \cdots [y_n]_\beta \right] \qquad \text{Repeated (4.3)}$$

Choose as bases the usual (good) orthonormal bases. For example, for a signal with four samples, define the bases by

$$\alpha = \beta = \left\{ \begin{bmatrix} 1 \\ 0 \\ 0 \\ 0 \end{bmatrix} \begin{bmatrix} 0 \\ 1 \\ 0 \\ 0 \end{bmatrix} \begin{bmatrix} 0 \\ 0 \\ 1 \\ 0 \end{bmatrix} \begin{bmatrix} 0 \\ 0 \\ 0 \\ 1 \end{bmatrix} \right\}$$

With this basis, the coordinates of both the input signal and the output transform are the input and output vectors. That is, step 2 is automatic. Now determine what the transform does to the input signal in order to find the matrix of transformation. To do this, consider Eqs. 10.2 and 10.3, the filter and downsample operations.

$$c_j(k) = \sum_m h_0(m - 2k) c_{j+1}(m) \qquad \text{Repeated (10.2)}$$

$$d_j(k) = \sum_m h_1(m - 2k) c_{j+1}(m) \qquad \text{Repeated (10.3)}$$

The initial values of $c_{j+1}(m)$ are the signal to be transformed. For the Haar wavelets, $h_0(m) = \left\{ 1/\sqrt{2} \ \ 1/\sqrt{2} \right\}$ and $h_1(m) = \left\{ 1/\sqrt{2} \ \ -1/\sqrt{2} \right\}$. These two operations find the sum and difference divided by $\sqrt{2}$. Suppose that there are $N = 2^j$ samples of a waveform $v(t)$. Apply the sum and difference transform to successive pairs of samples. There are $N/2$ or 2^{j-1} such pairs, giving rise to $N/2$ scaling function coefficients, and $N/2$ wavelet coefficients at level $j - 1$. Apply this procedure to the scaling coefficients at level $j-1$ to find the level $j-2$ coefficients, and so on until reaching level $j=0$.

In this paradigm there are N samples before transformation (or N coefficients c_{jk}.), and there are more than N coefficients after transformation. Use only N of these. Use c_{00} and all the wavelet coefficients d_{jk}. The number of coefficients is 1 for c_{00} plus 2^n for each d_n. This adds up to

$$1+\sum_{n=0}^{j-1} 2^n = 2^j$$

The filter h_0 is low-pass (the sum operation) and h_1 is high-pass (the difference operation). Downsampling by 2 selects the appropriate output coefficients. Note that the filters $h_0(-n)$ and $h_1(-n)$ are combined correlation and downsampling operations.

To illustrate, suppose that we start with four coefficients (or samples) $c_{2k} = [c_{20}\ c_{21}\ c_{22}\ c_{23}]$. The procedure is:

1. Sum and downsample to obtain $[c_{10}\ c_{11}]$.
2. Subtract and downsample to obtain $[d_{10}\ d_{11}]$.
3. Repeat using c_{1k} as input. Sum and downsample to obtain $[c_{00}]$.
4. Subtract and downsample to obtain $[d_{00}]$.

The four coefficients $[c_{00}\ d_{00}\ d_{10}\ d_{11}]$ form the wavelet transform of the original signal $[c_{20}\ c_{21}\ c_{22}\ c_{23}]$.

Example 10.5. For $N = 4$ there are four coefficients c_{jk} and four coefficients in the transform. The sum and difference operations are linear, meaning that this transform can be represented by a 4×4 matrix of transformation. Use the usual basis vectors and find the matrix of transformation.

Solution: Find the matrix A in the equation $y = Ax$, where

$$y = \begin{bmatrix} c_{00} \\ d_{00} \\ d_{10} \\ d_{11} \end{bmatrix} \qquad x = \begin{bmatrix} c_{20} \\ c_{21} \\ c_{22} \\ c_{23} \end{bmatrix}$$

That is, find A in

$$y = \begin{bmatrix} c_{00} \\ d_{00} \\ d_{10} \\ d_{11} \end{bmatrix} = \begin{bmatrix} a_{00} & a_{01} & a_{02} & a_{03} \\ a_{10} & a_{11} & a_{12} & a_{13} \\ a_{20} & a_{21} & a_{22} & a_{23} \\ a_{30} & a_{31} & a_{32} & a_{33} \end{bmatrix} \begin{bmatrix} c_{20} \\ c_{21} \\ c_{22} \\ c_{23} \end{bmatrix}$$

According to Eq. 10.2, the coefficient c_{00} is the sum of the c_{1k}'s, divided by $\sqrt{2}$:

$$c_{00} = \frac{c_{10} + c_{11}}{\sqrt{2}}$$

but

$$c_{10} = \frac{c_{20} + c_{21}}{\sqrt{2}} \tag{10.5}$$

and

$$c_{11} = \frac{c_{22} + c_{23}}{\sqrt{2}} \tag{10.6}$$

Therefore,

$$c_{00} = \frac{c_{20} + c_{21} + c_{22} + c_{23}}{2} \tag{10.7}$$

This gives the first row of A as [½ ½ ½ ½]. Use the same procedure to express d_{00} in terms of $\{c_{2k}\}$:

$$d_{00} = \frac{c_{10} - c_{11}}{\sqrt{2}}$$

Substituting Eqs. 10.5 and 10.6 gives

$$d_{00} = \frac{c_{20} + c_{21} - c_{22} - c_{23}}{2} \tag{10.8}$$

Use the same procedure to express d_{10} and d_{11} in terms of $\{c_{2k}\}$. This gives

$$d_{10} = \frac{c_{20} - c_{21}}{\sqrt{2}} \tag{10.9}$$

$$d_{11} = \frac{c_{22} - c_{23}}{\sqrt{2}} \tag{10.10}$$

Putting Eqs. 10.7 through 10.10 in matrix form gives

$$\begin{bmatrix} c_{00} \\ d_{00} \\ d_{10} \\ d_{11} \end{bmatrix} = \begin{bmatrix} \frac{1}{2} & \frac{1}{2} & \frac{1}{2} & \frac{1}{2} \\ \frac{1}{2} & \frac{1}{2} & -\frac{1}{2} & -\frac{1}{2} \\ \frac{1}{\sqrt{2}} & -\frac{1}{\sqrt{2}} & 0 & 0 \\ 0 & 0 & \frac{1}{\sqrt{2}} & -\frac{1}{\sqrt{2}} \end{bmatrix} \begin{bmatrix} c_{20} \\ c_{21} \\ c_{22} \\ c_{23} \end{bmatrix} \tag{10.11}$$

■

Drill 10.4. Find the inverse of the matrix above and show that it correctly derives the c_{2k} coefficients from the transform coefficients $[c_{00}\ d_{00}\ d_{10}\ d_{11}]$.

Answer:
$$A^{-1} = \begin{bmatrix} \frac{1}{2} & \frac{1}{2} & \frac{1}{\sqrt{2}} & 0 \\ \frac{1}{2} & \frac{1}{2} & -\frac{1}{\sqrt{2}} & 0 \\ \frac{1}{2} & -\frac{1}{2} & 0 & \frac{1}{\sqrt{2}} \\ \frac{1}{2} & -\frac{1}{2} & 0 & -\frac{1}{\sqrt{2}} \end{bmatrix}$$

Example 10.6. Find the matrix of transformation for a signal of eight samples.

Solution: Repeat the above procedure. Solving for c_{2k} and d_{2k} in terms of the c_{3k} coefficients gives

$$c_{20} = \frac{c_{30} + c_{31}}{\sqrt{2}} \qquad d_{20} = \frac{c_{30} - c_{31}}{\sqrt{2}}$$

$$c_{21} = \frac{c_{32} + c_{33}}{\sqrt{2}} \qquad d_{21} = \frac{c_{32} - c_{33}}{\sqrt{2}}$$

$$c_{22} = \frac{c_{34} + c_{35}}{\sqrt{2}} \qquad d_{22} = \frac{c_{34} - c_{35}}{\sqrt{2}}$$

$$c_{23} = \frac{c_{36} + c_{37}}{\sqrt{2}} \qquad d_{23} = \frac{c_{36} - c_{37}}{\sqrt{2}}$$

Substituting these into Eqs. 8.21 and 8.22 gives c_{00} and d_{00} in terms of c_{3k} (the first two rows) as

$$c_{00} = \frac{c_{30} + c_{31} + c_{32} + c_{33} + c_{34} + c_{35} + c_{36} + c_{37}}{2\sqrt{2}}$$

$$d_{00} = \frac{c_{30} + c_{31} + c_{32} + c_{33} - c_{34} - c_{35} - c_{36} - c_{37}}{2\sqrt{2}}$$

Also,

$$d_{10} = \frac{c_{20} - c_{21}}{\sqrt{2}} \qquad d_{11} = \frac{c_{22} - c_{23}}{\sqrt{2}}$$

Combining these with the equations above gives

$$d_{10} = \frac{c_{30} + c_{31} - c_{32} - c_{33}}{2} \qquad d_{11} = \frac{c_{34} + c_{35} - c_{36} - c_{37}}{2}$$

These give the third and fourth rows. Continuing on in this manner we find the matrix of transformation given by

$$A = \begin{bmatrix} \frac{1}{2\sqrt{2}} & \frac{1}{2\sqrt{2}} & \frac{1}{2\sqrt{2}} & \frac{1}{2\sqrt{2}} & \frac{1}{2\sqrt{2}} & \frac{1}{2\sqrt{2}} & \frac{1}{2\sqrt{2}} & \frac{1}{2\sqrt{2}} \\ \frac{1}{2\sqrt{2}} & \frac{1}{2\sqrt{2}} & \frac{1}{2\sqrt{2}} & \frac{1}{2\sqrt{2}} & -\frac{1}{2\sqrt{2}} & -\frac{1}{2\sqrt{2}} & -\frac{1}{2\sqrt{2}} & -\frac{1}{2\sqrt{2}} \\ \frac{1}{2} & \frac{1}{2} & -\frac{1}{2} & -\frac{1}{2} & 0 & 0 & 0 & 0 \\ 0 & 0 & 0 & 0 & \frac{1}{2} & \frac{1}{2} & -\frac{1}{2} & -\frac{1}{2} \\ \frac{1}{\sqrt{2}} & -\frac{1}{\sqrt{2}} & 0 & 0 & 0 & 0 & 0 & 0 \\ 0 & 0 & \frac{1}{\sqrt{2}} & -\frac{1}{\sqrt{2}} & 0 & 0 & 0 & 0 \\ 0 & 0 & 0 & 0 & \frac{1}{\sqrt{2}} & -\frac{1}{\sqrt{2}} & 0 & 0 \\ 0 & 0 & 0 & 0 & 0 & 0 & \frac{1}{\sqrt{2}} & -\frac{1}{\sqrt{2}} \end{bmatrix}$$

This matrix is *unitary* because $A^{-1} = A^t$.

■

Chapter 11
Using Wavelets

Wavelets are basis functions for the wavelet transform, just as exponentials $\{e^{j\omega t}\}$ are basis functions for the Fourier transform. Wavelet coefficients can be calculated in the same way as Fourier coefficients, by using the basis functions in an inner product calculation. However, wavelets allow us to use an alternative scheme involving samples of the waveform supplied to a filter-down sample operation. For this, we must have an appropriate filter $h_0(n)$. Different wavelets are associated with different filters.

This chapter broadens our view of wavelets by presenting the implementation of practical filters, an application to pattern recognition, and other practical examples. The application of practical wavelets is surprisingly simple, because it relies on filters and sample rate change.

Chapter Goals: After completing this chapter, you should be able to do the following:

- Given a FIR filter h_0, find the remaining three filters to complete a QMF bank.
- Compute the wavelet transform for a given signal.
- Reduce dimensionality in pattern recognition using wavelets.

11.1. Top-Down Approach

Method 1.

Section 8.2 classified wavelet transforms into three categories: the discrete wavelet transform (DWT), the continuous wavelet transform (CWT), and the transform for discrete-time signals. All use the same basic approach to calculate by correlation the wavelet transform for a given signal. Here are the steps to find the wavelet transform of a given signal.

1. Start at the beginning of the waveform and compare to a wavelet by correlation.
2. Shift the wavelet to the right and repeat step 1. Do this until you have covered the entire signal.
3. Scale (stretch) the wavelet and repeat steps 1 and 2.
4. Repeat steps 1 through 3 for all scales.

Figure 11.1 depicts this process. The signal is a noisy sine wave, and the two wavelets represent a short high-resolution wavelet and a longer low-resolution wavelet. For the CWT, the above process is implemented in the following way. Position the wavelet at the left side of the signal and calculate the correlation. Then move the wavelet ever so slightly to the right and repeat. Keep this up to get a continuum of correlation values for this one wavelet. Next, stretch the wavelet ever so slightly and repeat the correlation process at each position along the waveform. Repeating this stretching and correlating gives a continuum of correlation values along the scale axis. The continuous wavelet transform has a two-dimensional continuous domain. One axis represents shift and the other represents scale.

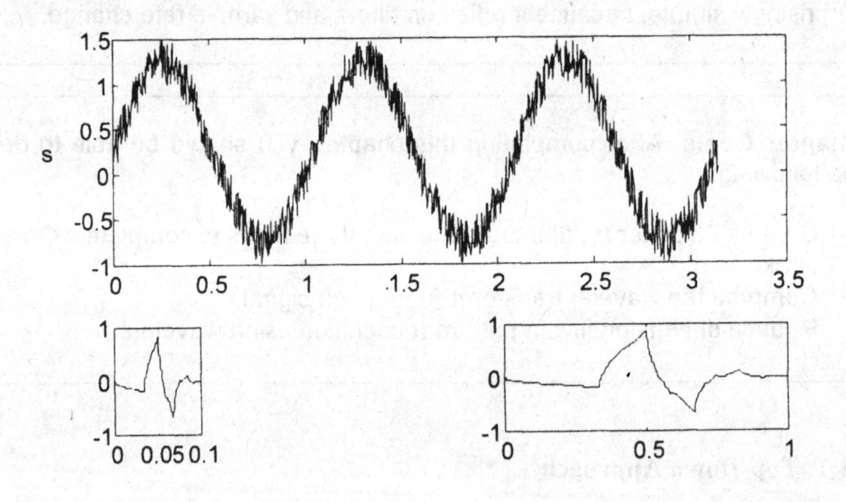

Figure 11.1. Waveform and two wavelets.

The DWT follows this general scheme in discrete steps. Shift the wavelet by discrete steps, and stretch the wavelet by discrete steps. This produces a two-dimensional discrete domain for the DWT. (Of course, the same can be said for the third transform for discrete-time signals since we begin with samples.) It turns out that this shifting and correlating can be done in discrete steps so that we lose no information with the DWT. However, in practice, method 2 is used.

Method 2.

Equations 10.2 and 10.3 give a relation from the scaling functions c_{j+1} to the next lower-level scaling and wavelet functions, c_j and d_j. This relation involves the filters h_0 and h_1. For given filters, these equations allow us to find the wavelet coefficients from samples of the signal, because the samples are the initial scaling function coefficients.

$$c_j(k) = \sum_m h_0(m-2k)c_{j+1}(m) \qquad \text{Repeated (10.2)}$$

$$d_j(k) = \sum_m h_1(m-2k)c_{j+1}(m) \qquad \text{Repeated (10.3)}$$

Figure 10.7, repeated here, depicts the general scheme for choosing scales and positions based on powers of 2, the *dyadic scales* and *positions*. This scheme applies to any-length filter, including Haar filters. Here is an example using db(2) filters to demonstrate how the DWT is applied in practice using method 2.

1. First, generate a noisy signal s of length 1024. (It is not necessary to start with a signal whose length is a power of 2, but it is more convenient for our purposes.)

```
N=1024;
t=linspace(0,pi,N);
s=sin(20.*t)+0.5*rand(1,N);
```

2. Define filters h_0 and h_1. [Use db(2) filters.]

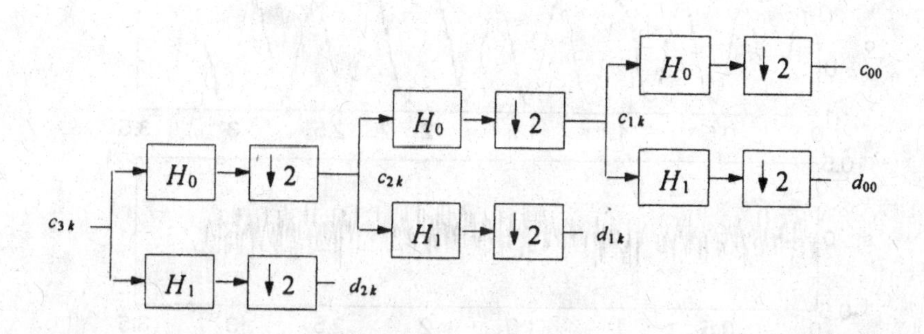

Repeated Figure 10.7

```
h0=[0.3415 0.5915 0.1585 -0.0915];
h1=[-0.0915 -0.1585 0.5915 -0.3415];
```

3. Filter and downsample the signal `s` to obtain the c and d sequences.

```
s0=conv(h0,s);
s1=conv(h1,s);
s0=s0(1,2:N+1); ← %Eliminate first value.
s1=s1(1,2:N+1); ←
s0=reshape(s0,2,N/2); ← downsample
s1=reshape(s1,2,N/2); ←
c=s0(1,:); ← The output c of length 512.
d=s1(1,:); ← The output d of length 512.
```

This produces the signals shown in Fig. 11.2. The top signal s contains 1024 samples, but the two derived signals c and d contain 512 samples due to downsampling. The signal s is the input to the first stage, labeled c_{3k} in Fig. 10.7, and the signals c and d are the output signals, labeled c_{2k} and d_{2k}. The signal d is the wavelet coefficients at this level. Following the numbering scheme in Fig. 10.7, these should be labeled d_{9k}, since d_{9k} has 512 values.

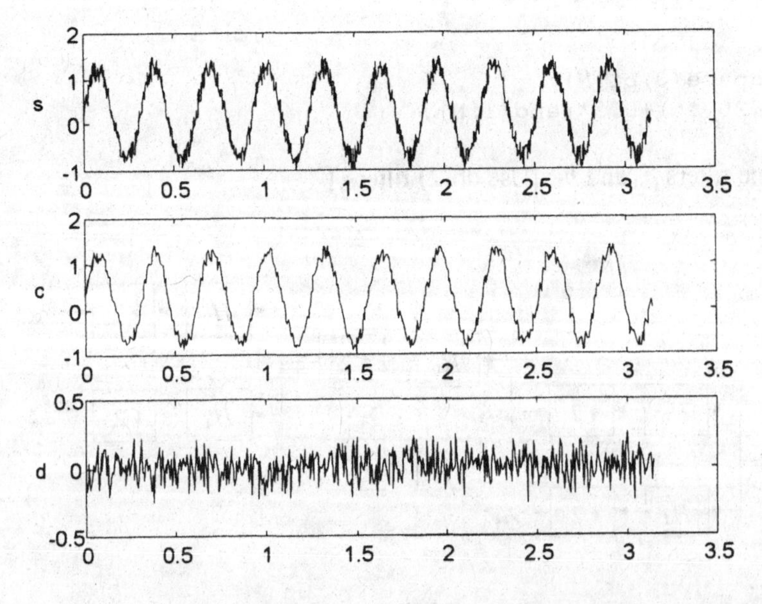

Figure 11.2. First iteration: signals `s`, `c`, `d`.

Drill 10.1. How many samples do d_{7k}, d_{5k}, and d_{3k} contain, respectively?

Answer: $2^7 = 128$, $2^5 = 32$, $2^3 = 8$.

Continuing this process according to Fig. 10.7, operate on the scaling function coefficients c_{9k} to produce c_{8k} and d_{8k}. This means repeating step 3 in the program above with $N = 512$ and $s = c_{9k}$. This produces the plot labeled "Second Iteration" in Fig. 11.3. The c and d signals now contain 256 points and should be labeled c_{8k} and d_{8k}.

Figure 11.4 shows the third iteration. Here, the c and d signals contain c_{7k} and d_{7k} with 128 samples each. Notice the general trend in this. The c signals become more like the sinusoidal part of the original signal, and each iteration extracts more and more noise from the signal. This illustrates one of the first and most obvious applications of wavelets to signal processing, *de-noising*. Do not assume from these diagrams that more and more processing will remove more and more noise. Figure 11.4 represents just about as much useful processing as possible on this signal. One or two more iterations and the c signal begins to look like noise.

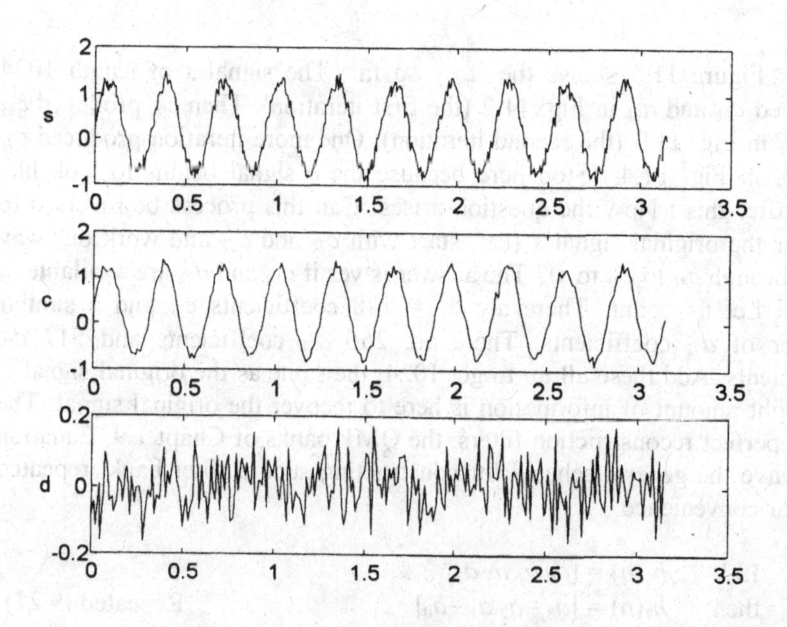

Figure 11.3. Second iteration.

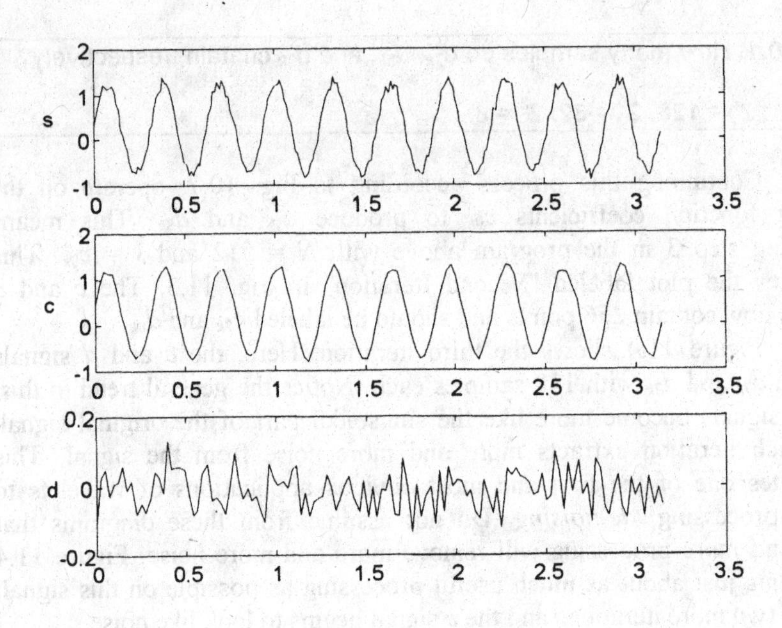

Figure 11.4. Third Iteration.

Figure 11.5 shows the story so far. The signal s of length 1024 produced c_{9k} and d_{9k} in Fig. 11.2 (the first iteration). Then c_{9k} produced c_{8k} and d_{8k} in Fig. 11.3 (the second iteration). One more iteration produced c_{7k} and d_{7k} in Fig. 11.4. (Stop here because the c signal begins to look like noise after this.) Now the question arises: Can this process be reversed to recover the original signal s (i.e., start with c_{7k} and d_{7k} and work our way back through c_{8k} to c_{9k} to s)? The answer is yes if d_{8k} and d_{9k} are available.

Let us count. There are $2^7 = 128$ coefficients c_{7k} and a similar number of d_{7k} coefficients. There are 256 d_{8k} coefficients and 512 d_{9k} coefficients. Add these all up to get 1024, the same as the original signal s. The right amount of information is here to recover the original signal. The key is perfect reconstruction filters, the QMF banks of Chapter 9. Equation 9.21 gave the general scheme for constructing such a filter bank, repeated here for convenience.

$$\begin{array}{ll} \text{If} & h_0(n) = [a_0\ a_1\ a_2\ a_3] \\ \text{then} & h_1(n) = [a_3\ {-a_2}\ a_1\ {-a_0}] \\ \text{and} & h_2(n) = [a_3\ a_2\ a_1\ a_0] \\ \text{and} & h_3(n) = [-a_0\ a_1\ {-a_2}\ a_3] \end{array} \qquad \text{Repeated (9.21)}$$

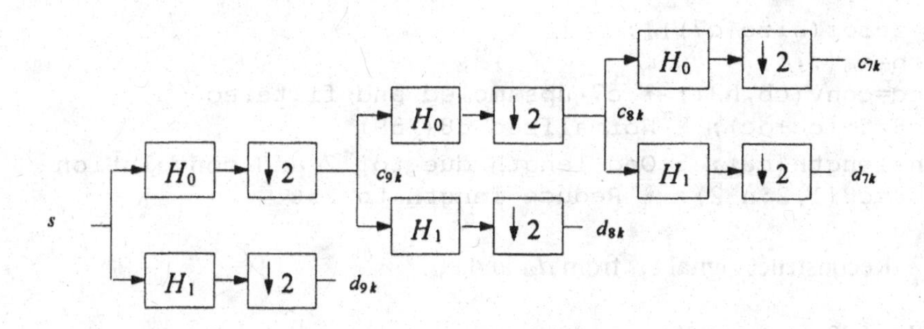

Figure 11.5. Picture of the processing in Figs. 11.2 through 11.4.

Now let us reconstruct the signal s from the wavelet transform. Figure 11.6 shows the procedure. Start with c_{7k} and d_{7k}, upsample, filter, and add to produce c_{8k}. Upsample and filter both c_{8k} and d_{8k}, then add to produce c_{9k}. Finally, upsample and filter both c_{9k} and d_{9k}, then add to produce s.

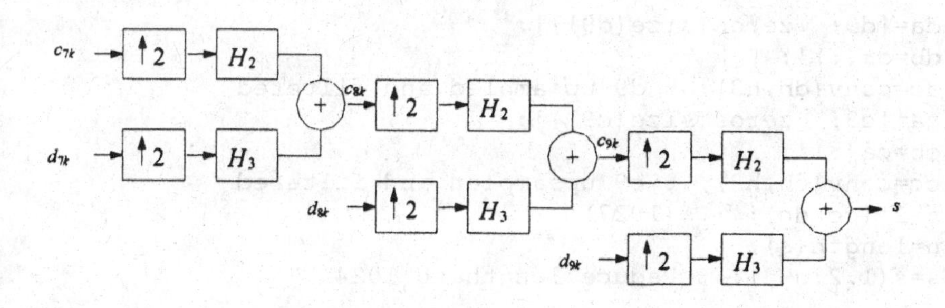

Figure 11.6. Inverse wavelet transform.

The following program segments are keyed to the various stages of the diagram in Fig. 11.6

1. Reconstruct signal c_{8k} from d_{7k} and c_{7k}.

```
da=[d7
 zeros(size(d7))]; %add row of zeros
db=da(:)';    %insert alternating zeros
dc=conv(db,h3); % d7 upsampled and filtered
ca=[c7
```

```
 zeros(size(c7))];
cb=ca(:)';
cc=conv(cb,h2); % c7 upsampled and filtered
c8=2*(cc+dc); % Normalized c8(259)
n=length(c8); % Odd length due to ↑2 and convolution
c8=c8(1,2:n-2); % Reduce length to 256
```

2. Reconstruct signal c_{9k} from d_{8k} and c_{8k}.

```
da=[d8;  zeros(size(d8))];
db=da(:)';
dc=conv(db,h3); % d8 upsampled and filtered
ca=[c8;  zeros(size(c8))];
cb=ca(:)';
cc=conv(cb,h2); % c8 upsampled and filtered
c9=2*(cc+dc); % Normalized c9(515)
n=length(c9);
c9=c9(1,2:n-2); % Reduce length to 512
```

3. Reconstruct signal s from d_{9k} and c_{9k}.

```
da=[d9;  zeros(size(d9))];
db=da(:)';
dc=conv(db,h3); % d9 upsampled and filtered
ca=[c9;  zeros(size(c9))];
cb=ca(:)';
cc=conv(cb,h2); % c9 upsampled and filtered
s=2*(cc+dc); % s(1027)
n=length(s);
s=s(1,2:n-2); % Reduce length to 1024
```

4. Plot signals s, c_{9k}, and d_{9k}. This produces Fig. 11.7.

```
t=linspace(0,pi,length(s));
subplot(3,1,1)
plot(t,s)
ylabel('s')
title('Reconstructed Signal')
t1=linspace(0,pi,length(c9));
subplot(3,1,2)
plot(t1,c9)
ylabel('c9')
subplot(3,1,3)
plot(t1,d9)
ylabel('d9')
```

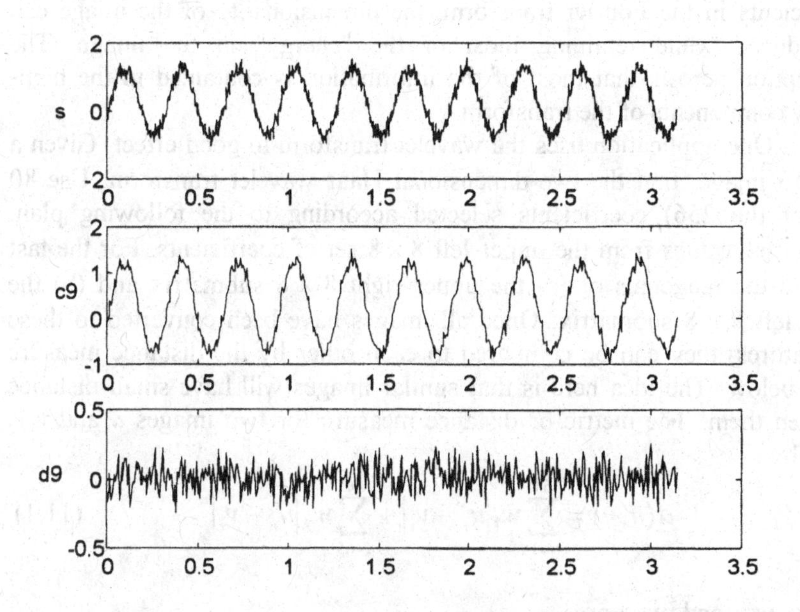

Figure 11.7. Reconstructed signal.

Since perfect reconstruction filters were used, this should be identical to Fig. 11.2. A comparison shows that the match is very good, differing only slightly in a few details. These differences are due to round-off and truncation error.

Consider the difference between the approach taken in Chapter 8 and the one taken here. Chapter 8 took what might be called a bottom-up approach, first calculating φ_{00} and ψ_{00}, followed by φ_{10}, φ_{11}, ψ_{10} and ψ_{11}, etc. This procedure starts at the other end. From 1024 values of the signal, calculate 512 values of φ_{9k} and ψ_{9k}, followed by 256 values of φ_{8k}, ψ_{8k}, etc. This is the top-down approach. Equations 10.2 and 10.3 are the formulas that allow us to do this.

11.2. Pattern Recognition

Humans can recognize objects in an image, but it is difficult for a machine to do so. The usual procedure is to extract features in the image and match these to known objects. These features can take various forms, from extracting lines and blobs to using some type of transform. The purpose of feature extraction is to reduce the size (dimensionality) of the image while retaining essential information. By using the largest (in magnitude) coefficients in the Fourier transform, the dimensionality of the image can be reduced while retaining most of the "energy" in the image. The assumption here is that most of the information is contained in the high-energy components of the transform.

One application uses the wavelet transform to good effect. Given a 16×16 image, find the two-dimensional Haar wavelet transform. Use 80 (out of the 256) coefficients selected according to the following plan. Extract 64 values from the upper left 8×8 set of coefficients. For the last 16, use the diagonals of (a) the upper right 8×8 submatrix and (b) the lower left 8×8 submatrix. Once all images have been converted to these 80 features, they can be compared to each other by the distance measure given below. The idea here is that similar images will have small distance between them. The metric or distance measure for two images u and v is given by

$$d(u,v) = \sum_{i=1}^{64} w_1 |u_i - v_i| + \sum_{i=65}^{80} w_2 |u_i - v_i| \tag{11.1}$$

where w_1 and w_2 are weighting coefficients that are chosen to optimize the process. Typically, $w_1 = 1$ and $w_2 = 1/4$.

Example 11.1. Let us apply the scheme above to the four images A, B, C, and D in Fig. 11.8. Find the distance between each pair of images using the first summation in Eq. 11.1 with $w_1 = 1$.

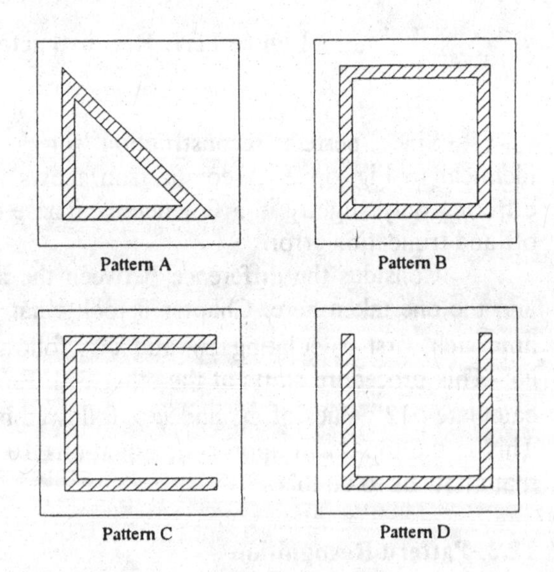

Figure 11.8. Patterns for Example 11.1.

Solution: To convert these images into MATLAB, define 16×16 matrices A, B, C, and D using 1's for the shaded areas and 0's for the blank areas of each image. For example, the A matrix is given by

```
A=[0 0 0 0 0 0 0 0 0 0 0 0 0 0 0 0
   0 0 0 0 0 0 0 0 0 0 0 0 0 0 0 0
   0 0 1 0 0 0 0 0 0 0 0 0 0 0 0 0
   0 0 1 1 0 0 0 0 0 0 0 0 0 0 0 0
   0 0 1 1 1 0 0 0 0 0 0 0 0 0 0 0
   0 0 1 1 1 1 0 0 0 0 0 0 0 0 0 0
   0 0 1 1 1 1 1 0 0 0 0 0 0 0 0 0
   0 0 1 1 0 1 1 1 0 0 0 0 0 0 0 0
   0 0 1 1 0 0 1 1 1 0 0 0 0 0 0 0
   0 0 1 1 0 0 0 1 1 1 0 0 0 0 0 0
   0 0 1 1 0 0 0 0 1 1 1 0 0 0 0 0
   0 0 1 1 0 0 0 0 0 1 1 1 0 0 0 0
   0 0 1 1 1 1 1 1 1 1 1 1 1 0 0 0
   0 0 1 1 1 1 1 1 1 1 1 1 1 1 0 0
   0 0 0 0 0 0 0 0 0 0 0 0 0 0 0 0
   0 0 0 0 0 0 0 0 0 0 0 0 0 0 0 0];
```

Matrices B, C, and D are defined similarly. Next, find the two-dimensional Haar transform. Since the dimension is small, we will use the matrix of transformation, selected according to the scheme outlined in Section 10.4. Let

$$p = \frac{1}{4} \qquad q = \frac{1}{2\sqrt{2}} \qquad r = \frac{1}{2} \qquad s = \frac{1}{\sqrt{2}}$$

Then the matrix of transformation W is given by

```
W=[p p p p p p p p p p p p p p p p
   p p p p p p p p -p -p -p -p -p -p -p -p
   q q q q -q -q -q -q 0 0 0 0 0 0 0 0
   0 0 0 0 0 0 0 0 q q q q -q -q -q -q
   r r -r -r 0 0 0 0 0 0 0 0 0 0 0 0
   0 0 0 0 r r -r -r 0 0 0 0 0 0 0 0
   0 0 0 0 0 0 0 0 r r -r -r 0 0 0 0
   0 0 0 0 0 0 0 0 0 0 0 0 r r -r -r
   s -s 0 0 0 0 0 0 0 0 0 0 0 0 0 0
   0 0 s -s 0 0 0 0 0 0 0 0 0 0 0 0
   0 0 0 0 s -s 0 0 0 0 0 0 0 0 0 0
   0 0 0 0 0 0 s -s 0 0 0 0 0 0 0 0
   0 0 0 0 0 0 0 0 s -s 0 0 0 0 0 0
   0 0 0 0 0 0 0 0 0 0 s -s 0 0 0 0
   0 0 0 0 0 0 0 0 0 0 0 0 s -s 0 0
   0 0 0 0 0 0 0 0 0 0 0 0 0 0 s -s];
```

To find column transforms, multiply W times A to get

$$y = W * A \tag{11.2a}$$

Then find the row transform by

$$x = W * y^t \tag{11.2b}$$

This completes the two-dimensional transform operation. Next, we select the upper left 8×8 submatrix for each pattern and compare distances. Using Eq. 11.1 with $w_1 = 1$, $w_2 = 0$, we get

$$\begin{array}{ll} d(A,B) = 38.97 & d(A,C) = 31.74 \\ d(A,D) = 31.74 & d(B,C) = 13.07 \\ d(B,D) = 13.07 & d(C,D) = 22.14 \end{array}$$

These distances "look" to be correct. Note that $d(A,C) = d(A,D)$, as we would expect from an inspection of the patterns. Also, $d(B,C) = d(B,D)$. We would expect these results if we measured distance between the original 64×64 images. The fact that we can use only 64 transformed pixels from each image to measure distance represents a considerable saving in machine time. This scheme works well for larger images, say 128×128, using less than 100 transformed pixels for the comparison. This allows us to search a large database for a particular image in reasonable time.

■

11.3. Hidden Singularities

Wavelets can detect hidden information in the derivatives of functions. Here is an example, due to Wayne Galli and Ole Nielson, where the wavelet decomposition detects a feature that cannot be seen directly from the original signal. Consider the function

$$f(t) = \begin{cases} t^3/6 & 0 \le t < \frac{1}{2} \\ t^3/6 - t^2/2 + t/2 - 1/8 & 0 \le \frac{1}{2} < 1 \end{cases}$$

The first derivative of f is

$$f'(t) = \begin{cases} t^2/2 & 0 \le t < \frac{1}{2} \\ t^2/2 - t + 1/2 & 0 \le \frac{1}{2} < 1 \end{cases}$$

and the second derivative is

$$f'(t) = \begin{cases} t & 0 \le t < \frac{1}{2} \\ t - 1 & 0 \le \frac{1}{2} < 1 \end{cases}$$

Figure 11.9 shows a plot of 1024 sampled function values of these three functions. The function is continuous and smooth at t = ½. The first derivative is continuous, and the second derivative is discontinuous.

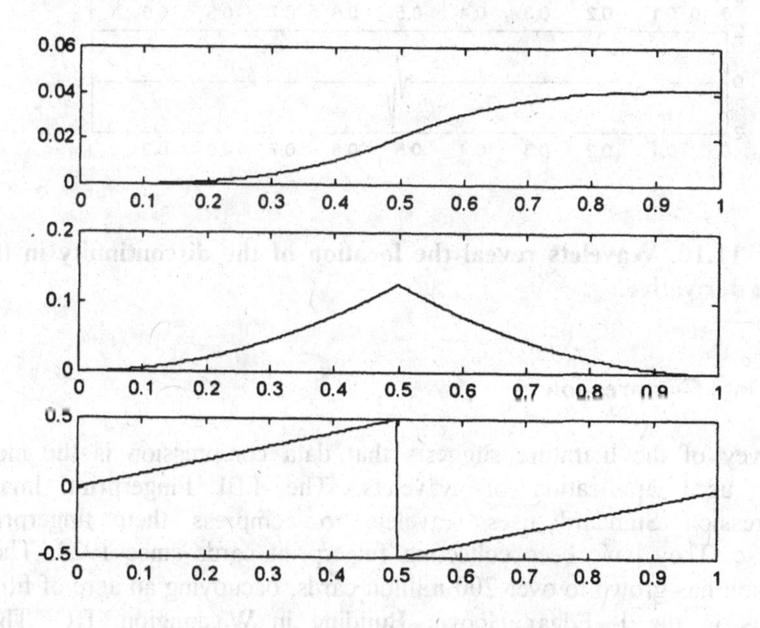

Figure 11.9. The function *f*(*t*) and its derivatives.

Figure 11.10 shows the wavelet decomposition. The top diagram is the original function. The succeeding diagrams show d_9 (512 samples), d_8 (256 samples), and d_7 (128 samples). The db(4) wavelet with eight coefficients was used in this example. The discontinuity in the second derivative causes the activity in the wavelets at t = ½.

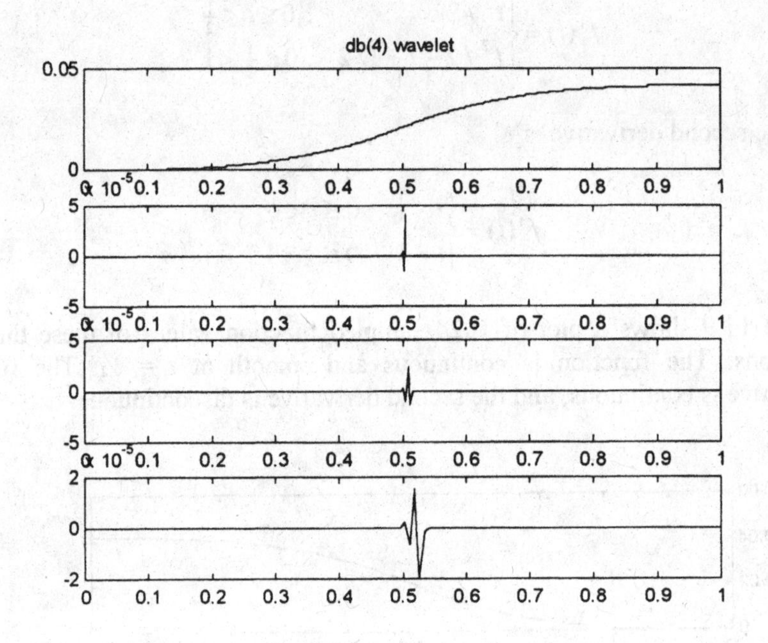

Figure 11.10. Wavelets reveal the location of the discontinuity in the second derivative.

11.4. Data Compression

A survey of the literature suggests that data compression is the most widely used application of wavelets. The FBI Fingerprint Image Compression Standard uses wavelets to compress their fingerprint database. They have been collecting fingerprint cards since 1924. Their collection has grown to over 200 million cards, occupying an acre of filing cabinets in the J. Edgar Hoover Building in Washington, DC. They examine some 29 million records each time they are asked to match a print. They need help.

The FBI digitizes fingerprints at 500 dots per inch with 8 bits of gray-scale resolution. One fingerprint card (10 fingers) contains about 10 MB of data. A 32-kbit modem would require more than 30 minutes to send one card. Some form of data compression is needed, and the first candidate is lossless compression. But in practice, lossless methods can produce no more than 2:1 data compression. Lossy methods become attractive, and this is where wavelets come into the picture.

The new ISO JPEG standard will work on any type of image, but experiments show that it deletes some important details and imposes a "blocky" pattern on the image. The FBI's fingerprint experts prefer the wavelet image because it leaves some of the details intact and does not impose a blocky look on the image. It tends to "blur" rather than "block," and this is not as offensive to the user.

The basic idea behind lossy data compression is to delete the low-energy components of an image (or signal) and retain the high-energy data, or else to delete the high-frequency data and retain the low-frequency data, or both. The application determines which combination is best, and wavelets make it easy to choose. Here are the results of an experiment made on the pattern recognition data in Fig. 11.8.

Figure 11.11 shows the pattern A and its compressed transform. The idea is to use Eqs. 11.2a and 11.2b to find the forward transform and zero each quadrant except the upper left. This compresses the image into an 8×8 matrix, thus reducing the data by a 4:1 ratio. To see what damage this has done to the original image, the inverse is generated by the following operations.

The wavelet matrix of transformation is unitary, meaning that the inverse is the transpose. That is,

$$W^{-1} = W^{t}$$

If x is the 16×16 transform of A (from Eq. 11.2b) then the inverse is given by

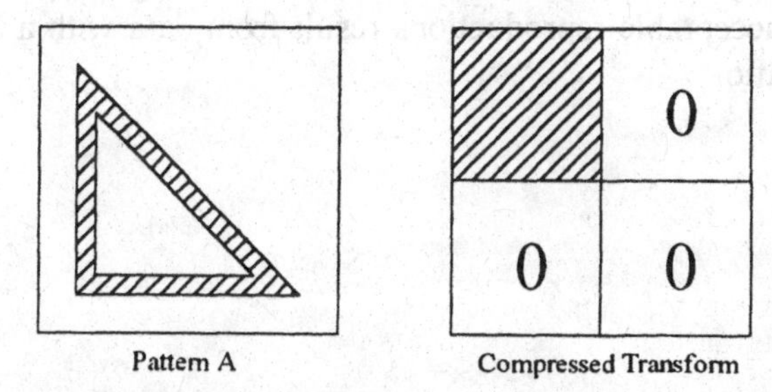

Figure 11.11. To compress the 16 × 16 image A, leave the upper left quadrant of the transform unchanged and zero the other three quadrants.

$$y = W^t * x \tag{11.3a}$$

$$A = W^t * y^t \tag{11.3b}$$

These operations are just the inverse of Eqs. 11.2. Let the symbol p stand for 0.75. Then the resulting image is given by

```
A=[0 0 0 0 0 0 0 0 0 0 0 0 0 0 0 0
   0 0 0 0 0 0 0 0 0 0 0 0 0 0 0 0
   0 0 p p 0 0 0 0 0 0 0 0 0 0 0 0
   0 0 p p 0 0 0 0 0 0 0 0 0 0 0 0
   0 0 1 1 p p 0 0 0 0 0 0 0 0 0 0
   0 0 1 1 p p 0 0 0 0 0 0 0 0 0 0
   0 0 1 1 p p p p 0 0 0 0 0 0 0 0
   0 0 1 1 p p p p 0 0 0 0 0 0 0 0
   0 0 1 1 0 0 0 p p p 0 0 0 0 0 0
   0 0 1 1 0 0 0 p p p 0 0 0 0 0 0
   0 0 1 1 0 0 0 0 p p p p 0 0 0 0
   0 0 1 1 0 0 0 0 p p p p 0 0 0 0
   0 0 1 1 1 1 1 1 1 1 1 1 p p 0 0
   0 0 1 1 1 1 1 1 1 1 1 1 p p 0 0
   0 0 0 0 0 0 0 0 0 0 0 0 0 0 0 0
   0 0 0 0 0 0 0 0 0 0 0 0 0 0 0 0];
```

The diagonal values are reduced in magnitude and spread by one or two pixels. Otherwise, the image is unchanged.

This crude example shows the type of distortion that appears in compressed images in a simple scheme. By fine tuning the procedure on a larger image, acceptable reproductions result from data with a 9:1 or more compression ratio.

Index

aliasing, 82
allpass networks, 170ff, 174
Argand diagram, 99

Banach space, 25
basis, 51ff
Beagle, the, 16
biorthogonality, 210ff
Boltzmann, Ludwig, 17
Brown, Robert, 16
Brownian motion, 17

CBS inequality, 45
codomain, 5
companding, 93ff
conjugate QMF, 182
convolution, 116
coordinates, 53, 66
correlation, 110

data compression, 232ff
Daubechies, Ingrid, 192
Daubechies filters, 208
Darwin, Charles, 16
decimation, 95
deterministic signal, 16ff
dimension, 54
direct sum, 151
discrete metric, 32
discrete wavelet transform, 162ff
domain, 5
dot product space, 31
downsampling, 95ff

Einstein, Albert, 17
energy, 10
energy signal, 9ff
entropy, 17

FFT, 121ff
FFT as downsampling, 139ff
FFT as matrix decomposition, 125ff
fields, 7ff
filter banks, 186ff
finite impulse response, 170
Fourier transforms, 18ff
fractional rate change, 104ff
function, 5

Gali, Wayne, 230
geometric series, 15

Haar functions, 149
Haar transforms, 21ff
Hermitian, 40
hidden singularities, 230ff
high-pass filter, 173
Hilbert space, 25

independence, 48
inner product, 39ff
inner product space, 31
interpolation, 95

Johnson noise, 18

linear independence, 48ff
linear phase, 184
low-pass filter, 171

magic part, 201ff
map, 6
mapping, 6
matrix (definition), 69
matrix of transformation, 70, 213
maximum phase, 180
metric space, 31ff
minimum phase, 179
mirror image, 178
multiresolution analysis, 167

Nielson, Ole, 230
nonperiodic sampling, 88ff
norm, 36ff
Nyquist rate, 80ff

orthogonal complement, 150
orthogonality, 45ff

pattern recognition, 227ff
perfect reconstruction, 186
power, 9
power signal, 9
pulse code modulation, 90ff

QMFs, 181ff
quantization, 90ff
quantization noise, 91

random signals, 16ff
random variable, 25
range, 6
rank, 51
reciprocal basis, 56ff
rms value, 11

sampling, 80ff
scaling function, 21
signal classification, 9
signal flow graph, 121
span, 54
subspace, 29

themal noise, 17
transforms, 54ff
two-scale relationship, 167

upsampling, 95, 103ff

vector space, 27
vectors, 25ff
Von Neumann, John, 25

wavelet, 21
wavelet transform, 1, 154ff